☑ **W9-DJE-125**

WITHDRAWN

Contributions of the Committee on Desert and Arid Zones Research
of the Southwestern and Rocky Mountain Division of the
American Association for the Advancement of Science

Previous Symposia of the Series

The Reclamation of Disturbed Arid Lands

The Committee on Desert and Arid Zones Research
of the Southwestern and Rocky Mountain Division
of the
American Association for the Advancement
of Science

Statement of Purpose

The objective of the Committee on Desert and Arid Zones Research is to encourage the study of phenomena relating to and affected by human occupation of arid and semiarid regions, primarily within the areas represented by the Southwestern and Rocky Mountain Division of the A.A.A.S. This goal involves educational and research activities, both fundamental and applied, that may further understanding and efficient use of our arid lands.

COMMITTEE

Chairman:
    Marvin L. Riedesel, University of New Mexico, Albuquerque
Secretary:
    David K. Northington, Texas Tech University, Lubbock
Members:
    Joseph R. Goodin, Texas Tech University, Lubbock
    John A. Ludwig, New Mexico State University, Las Cruces
    Vern McMahon, University of Wyoming, Laramie
    James W. O'Leary, University of Arizona, Tucson
    Klaus D. Timmerhaus, University of Colorado, Boulder
    Robert A. Wright, West Texas State University, Canyon

Mailing Address
Dr. Max Dunford, Executive Secretary
Southwestern and Rocky Mountain Division of A.A.A.S.
P.O. Box 3 AF
Las Cruces, New Mexico 88001

The Reclamation of Disturbed Arid Lands

Edited by

Robert A. Wright

UNIVERSITY OF NEW MEXICO PRESS

Albuquerque

Library of Congress Cataloging in Publication Data

Main entry under title:
The Reclamation of disturbed arid lands.

(Contribution of the Committee on Desert and Arid Zones Research of the Southwestern and Rocky Mountain Division of the American Association for the Advancement of Science; 17)

"The Committee on Desert and Arid Zones Research . . . sponsored a symposium held in Denver, Colorado, February 23–24, 1977."

Includes bibliographies.

1. Reclamation of land—The West—Congresses. 2. Strip mining—Environmental aspects—The West—Congresses. 3. Arid regions—The West—Congresses. 4. Revegetation—The West—Congresses. I. Wright, Robert A., 1933–   II. American Association for the Advancement of Science. Southwestern and Rocky Mountain Division. Committee on Desert and Arid Zones Research. III. Series: American Association for the Advancement of Science. Southwestern and Rocky Mountain Division. Committee on Desert and Arid Zones Research. Contribution; 17.

S621.5.S8R44   631.6   78-1822
ISBN 0-8263-0476-1

Preface

Robert A. Wright
West Texas State University

There is an increasing demand by society for the reclamation of lands disturbed by man's activities. In response to this demand, the Committee on Desert and Arid Zones Research of the Southwestern and Rocky Mountain Division of the American Association for the Advancement of Science sponsored a symposium held in Denver, Colorado, February 23–24, 1977. The symposium focused on current and recent research on the reclamation of lands disturbed primarily by mining activity in the arid United States (*arid United States* being taken to include those areas with mean moisture deficiency).

This volume is divided into three parts, reflecting the three sessions of the symposium. Part 1 provides an overview of research on the reclamation of mined lands; part 2 is concerned with several phases of reclamation and related studies being carried on at Argonne National Laboratory; part 3 provides an in-depth look at specific projects concerned with various aspects of reclamation research in a number of areas of the arid United States.

These papers constitute a fairly representative sample of the enormous amount of work going on in the reclamation of disturbed lands. An environmentalist attending the meetings complained, however, that something important was lacking in the presentations: a love of nature. I believe that this indictment is only superficially true, in that such a feeling was not explicit. But we must look beyond this complaint. Disruption of nature is now and will continue to be a fact of life. The nature lover's cry of protest may be necessary, but it is not sufficient. To minimize damage and to maintain stewardship of the land, we must proceed from a rigorous, scientifically sound basis. It is this substantive aspect of reclamation that concerns the authors of these papers.

Contents

Part 1
An Overview of Research

1

Reestablishment of Woody Plants on Mine Spoils and Management of Mine Water Impoundments: An Overview of Forest Service Research on the Northern High Plains[1]

Ardell J. Bjugstad

Forest Research Laboratory
Rocky Mountain Forest and Range Experiment Station
South Dakota School of Mines and Technology
Rapid City, South Dakota

The northern High Plains region—that portion of the Great Plains west of the Missouri River in North Dakota, South Dakota, and extending into Montana and Wyoming—is noted for its wide expanses of flat to rolling grasslands and shrub steppe interspersed with drainageways and associated woody draws. This 65-million-acre region is highly valued for its wildlife habitats—including aquatic habitats—and stable watersheds as well as for its minerals and agronomic uses. The wooded drainageways also protect domestic livestock and produce fuel wood.

The climate of the High Plains is exceptionally variable. Occasional severe winters cause tremendous losses of both livestock and wildlife. One 1975 blizzard caused livestock losses of more than $2 million in South Dakota alone. Those losses would have been substantially greater without the protection of shrubs and trees in wooded draws and shelterbelts.

The woody draws, a critical habitat for many wildlife species, are the *only habitat* for several species. They provide 70 percent of the habitat for sharptail grouse and approximately 50 percent of the habitat for plains deer. Consequently, if woody draws are allowed to deteriorate and/or are not perpetuated in the plains ecosystem, deer and sharptail grouse populations could decrease drastically.

Inventories indicate that the woody draws (elm-ash-cottonwood type with associated oak and juniper) occupy approximately 650,000 acres (1 percent) of the 65-million-acre region. The removal of old, low-vigor trees to induce regeneration in a 70-year-old stand produced three and one-half cords of wood per acre in pieces with diameter larger than three inches. If 5 percent of the woody draw area that now exists were so treated each year, the gross return of energy could be worth an estimated $2.8 million. Consequently, the existing woody draws could produce a tangible annual income of at least $3 million based on potential wood and game production alone. Other benefits, such as protection, aesthetic values, water, and wind management, are extras.

The preceding figures reveal the relative smallness of the area occupied by the woody draws today in the High Plains, but at the same time emphasize the potential importance of the draws—much greater than their size in proportion to the total area. It is that importance that has provided the mandate for the Rocky Mountain Forest and Range Experiment Station to initiate research on the maintenance and reestablishment of woody species in the High Plains, particularly for areas drastically disturbed by mining. Although only a small percentage of the High Plains—about 0.6 percent in Wyoming—will be drastically disturbed by surface mining for coal (Thilenius and Glass 1974), the value of the wooded areas accentuates the need for new and innovative research.

THE ROCKY MOUNTAIN FOREST AND RANGE EXPERIMENT STATION

The Rocky Mountain Forest and Range Experiment Station has its headquarters at Fort Collins, Colorado, and field laboratories at Fort Collins; Bottineau, North Dakota; Rapid City, South Dakota; Lincoln, Nebraska; Laramie, Wyoming; Albuquerque, New Mexico; and Flagstaff and Tempe, Arizona (Fig. 1.1). The field laboratories at Rapid City, Laramie, and Albuquerque are researching the reclamation of surface mine spoils. Of those three, the work of the research work units at Rapid City and Laramie is discussed here.

The Rapid City Unit

At the Research Work Unit at Rapid City, six scientists are involved on a part-time basis with research on the reclamation areas surface mined for coal and bentonite clay. Included are two hydrologists, a range scientist, a wildlife biologist, a geologist, and an aquatic biologist. The unit's assignment is to "develop techniques for revegetating mine spoils and associated waters to enhance habitat for non-game birds, deer, small mammals, and waterfowl." More explicitly, the unit's mission is to conduct research to provide guidelines for the reestablishment of shrubs and trees in depressions and draws that are characteristic of the High Plains, and for

Figure 1.1   Rocky Mountain Forest and Range Experiment Stations

the mitigation of possible detrimental effects of surface mining on groundwater and surface water. The Rapid City Research Work Unit concentrates its efforts in central and northeastern Wyoming, northwestern South Dakota, and southwestern North Dakota.

The Laramie Unit

The Research Work Unit at Laramie, Wyoming, with one full-time scientist, is part of the Surface Environment and Mining (SEAM) Program. The purpose of the research is to determine and classify the ecosystems of the Eastern Powder River Coal Basin. The program comprises three parts: (1) development of an ecosystem classification of the area using the ECOCLASS hierarchy; (2) evaluation of the currently available remote-sensing technology to determine whether it is possible to map ecosystems defined at the *Habitat Type* (or *Community Type*) levels of the ECOCLASS hierarchy; and (3) determination of the primary production and abiotic regime of major ecosystems as a basis for defining their site potential.

The Laramie and Rapid City research units, then, have complementary functions. One determines ecological relationships, while the other develops techniques to rehabilitate the vegetation types based on the ecological relationships. That functional relationship becomes more clear after a review of the current and recent work of these units.

THE LAND AREA

Description

The land area of major concern is the Northern Great Plains coal province of the Fort Union formation. The area encompasses the Powder River Basin (Fig. 1.2), the southern extension of the Fort Union formation, and the Williston Basin (Packer 1974). The Powder River Basin extends from southern Montana into northeastern Wyoming; the Williston Basin includes eastern Montana and all of western North Dakota. The Powder River Basin contains a reserve of nearly 240 billion tons of subbituminous coal (Glass 1972), and the Williston Basin about 440 billion tons of lignite (Landis 1973). The other land area of concern is the bentonite clay beds east of the Powder River Basin in Wyoming and Montana and west and northwest of the Black Hills.

According to Yamamoto (1975), the climate of the two basins of the Fort Union formation in the northern Great Plains is continental, temperate, and semiarid. Average precipitation varies slightly between the basins. The average annual precipitation in the Powder River Basin is 14 inches, with

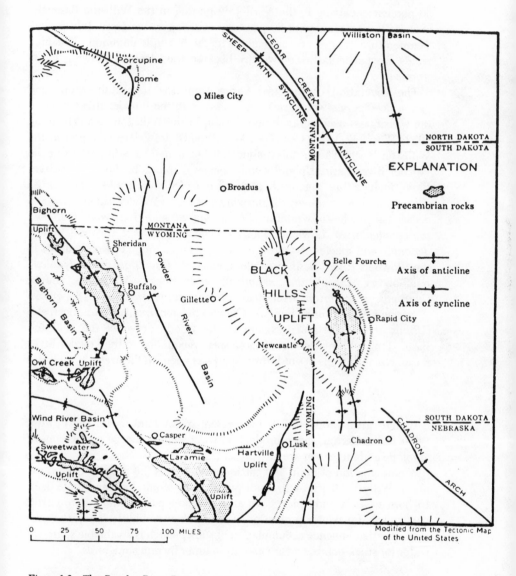

Figure 1.2 The Powder River Basin in Relation to Nearby Structural Features (Source: Final Environmental Impact Statement. Eastern Powder River Coal Basin of Wyoming, USDA, ICC, USDI 1974)

50 percent occurring in the April–July period; in the Williston Basin the average is 16 inches, with 60 percent occurring in the April–July period. Temperature ranges are extreme, with high winds common. The high winds are particularly noteworthy because they lead to blowing snow-storms.

The vegetation of the area varies from the sagebrush steppe and grama-needlegrass-wheatgrass type common in the Powder River Basin to the wheatgrass-needlegrass type common in the Williston Basin (Jameson 1952; Woolfolk 1949; Costello 1944). Woody draws are common in the Williston Basin. They are less abundant but nonetheless important in the Powder River Basin. Upland woody species common to the Powder River Basin include big sagebrush (*Artemisia tridentata*), rubber rabbitbrush (*Chrysothamnus nauseosus*), fourwing saltbush (*Atriplex canescens*), and occasional patches of winterfat (*Eurotia lanata*) and greasewood (*Sarcobatus vermiculatus*). Lowland woody species include cottonwood (*Populus deltoides*) and species of willow (*Salix*). Species common to woody draws in the Powder River Basin include green ash (*Fraxinus pennsylvanica*), buffaloberry (*Shepherdia argentea*), and Rocky Mountain juniper (*Juniperus scopulorum*). American plum (*Prunus americana*), hawthorn (*Crataegus rotundifolia*), bur oak (*Quercus macrocarpa*), and American elm (*Ulmus americana*) are prevalent in the woody draws of the Williston Basin along with scatterings of those species common to the Powder River Basin. Numerous species of forbs and grasses occur in the area.

Land Use

Range cattle production is the present major land use in the Powder River Basin. Earlier, sheep production was common. Large populations of antelope and mule deer are also present.

A combination of grain and cattle production is the prevailing land use in the Williston Basin, with small populations of antelope and mule deer. In both cases, woody plants are an important part of the environment, particularly for winter protection of livestock and wildlife. They also provide sites that enhance accumulation of snow, which is a valuable source of water for stock ponds, and are used in summer by nongame birds.

THE PROBLEMS

Mining and processing of the gigantic surface-minable reserves of lignite and subbituminous coal and bentonite clay in the High Plains area will no doubt disturb the area drastically. Some of the major mining-related dangers are that (1) altered land drainage patterns could result in large salt playas, (2) sporadic natural regeneration of woody plant species could make natural reestablishment difficult or impossible under some conditions, (3) the possible high cost of artificially establishing various woody plant species because of the lack of knowledge about specific site requirements might discourage efforts, and (4) the amelioration of harsh sites, including additional water, might be essential to enhance their rehabilitation.

RESEARCH

The importance of woody species for production of browse, protection of animals and soil, snow accumulation, stream-channel stability, and habitat suggested the research "to develop guidelines for the establishment or reestablishment of woody species adaptable to draws and upland sites."

Regional Research

The first effort needed was a premining inventory of drainage systems and other topographical features, including a determination of the magnitude of first- and second-order streams. Such information would provide guidelines for returning an area to a relatively mature premining topography. Preliminary interpretation of the data for the Powder River Basin suggests that variations in topographic variables must be interpreted on a regional scale as well as on a local scale, because disruptions from surface mining will affect not only the visual aesthetics, but also associated land resources of neighboring areas.

Other studies with regional aspects pertain to the determination of the effects of surface coal mining on the shallow groundwater in the eastern part of the Powder River Basin, and the effects of air pollution on lichens and ecological succession. The emphasis in groundwater research is to determine whether surface coal mining will affect local water wells. Again, preliminary interpretation of the data suggests that surface mining could affect shallow groundwater aquifers up to a quarter of a mile from the mine site.

During preliminary work in the Eastern Powder River Coal Basin, the Laramie Unit discovered the existence of an extremely well-developed lichen flora. Lichens are very sensitive to SO_2 pollution. Much of the coal mining will take place immediately upwind of the area of best lichen development. A study of the baseline ecology of lichen ecosystems in this area was started in June 1975. First emphasis was given to the taxonomy of lichens—a very specialized subject. This research was done through a contract with the University of Colorado at Boulder. The second phase of the study will be to initiate regular periodic monitoring of the lichens of selected sites to determine normal changes in their composition and growth. The final phase will be to subject specific lichens to known concentrations of SO_2 under laboratory conditions to determine their responses to this pollutant.

Site-Specific Research

Other research at the Rapid City Research Unit pertains to rehabilitation of coal and bentonite mine spoils, and management of mine water impoundments. Rehabilitation research focuses on the reestablishment of shrubs, trees, and forbs, while the water research concentrates on water quality for waterfowl, aquatic plants, and aquatic invertebrate animals.

Reclamation of Mine Spoils: Research shows that dryland techniques for reestablishment of shrub and tree species—such as green ash, Russian olive, ponderosa pine, Rocky Mountain juniper, silver buffaloberry, American plum, and Siberian peashrub—on bentonite and low-salt coal spoils in northeastern Wyoming have been moderately successful (Orr 1975). Green ash has the highest survival rate, 44 percent at the start of the third growing season. Rocky Mountain juniper and Russian olive are next with 24 percent; buffaloberry, Siberian peashrub, ponderosa pine, and American plum show survival rates of 7 percent or less.

Recently, the use of drip irrigation and greenhouse-grown, containerized planting stock has brought encouraging results. During the first winter, losses for irrigated plants averaged 16 percent, compared to 29 percent for nonirrigated. However, differences in results among species were striking. American plum had a winter mortality loss of only 6 percent under irrigation but 29 percent without irrigation. The container-grown stock survived better than bare root stock.

Several problems have developed in the low-maintenance irrigation systems: ruptured tubing, rodent damage, and clogged water emitters and filters. The systems need modification and improvement.

Other research in progress involves mixing mine overburden materials to achieve a mixture of sand, silt, and clay that resembles desirable soil. The soil that resulted from the spoil mix study is being used in research on the reestablishment of upland shrubs—big sagebrush, rubber rabbitbush, winterfat, fourwing saltbush, and Rocky Mountain juniper—and the reestablishment of various grass mixtures.

The cultural treatments for reestablishment of grass included straw and fiber mulch, fertilizer, and overhead irrigation. The shrub treatments involved the use of fertilizer, drip irrigation, and pine wood chips as a mulch. A good distribution of precipitation during the summer of 1976 in Wyoming resulted in good initial stands of grass in all treatments. With irrigation, the wood chip mulch yielded the highest survival rate for shrubs the first year. A greenhouse study is now under way to determine the possible effects of the leachate from pine wood chips on container-grown grass plants.

Other research is under way on: ways to manage utilization of shrubs and trees by antelope (*Antelocopra americana*), deer (*Odocoileus* sp) white-tailed jackrabbits (*Lepus*), and cottontail rabbits (*Sylvilagus*); how non-game birds are associated with reclaimed mine spoils and haul roads; and how to establish perennial forbs on coal and bentonite mine spoils.

Rehabilitation and Management of Impounded Mine Water: Research to develop guidelines for management of mine water impoundments was started by the Rapid City Unit in the spring of 1976. The assignment is to assess how the quality of impounded water in strip-mined areas relates to habitats of wildlife, including aquatic organisms.

Aquatic invertebrates are being collected from coal and bentonite strip mine ponds in central and southwestern North Dakota, northwestern South Dakota, and northeastern Wyoming. The species, diversity, and abundance of invertebrates determines the pond's capacity to meet the nutritional requirements of breeding and migratory waterfowl. Preliminary observations indicate that planktonic invertebrates are much more abundant in older ponds.

Chemical and physical characteristics of pond water are being investigated in North Dakota, South Dakota, and Wyoming. Water quality determines the kinds and abundance of aquatic plants and animals that can exist in the ponds that develop after strip mining. Results to date indicate that high salinity and turbidity are the major water quality problems.

Basin morphometry will be developed for the new ponds in the Dakotas and Wyoming. Slope and area of the basin surrounding a pond will help in determining the potential sediment load to the pond from erosion. Knowledge of pond bottom morphometry will help in determining how many shallow areas exist. Aquatic vegetation, which provides waterfowl cover and supports production of invertebrate organisms, grows in shallow areas. Analysis of pond morphometry will also indicate how much deep open water will be present for waterfowl.

Waterfowl studies are planned for ponds in the Dakotas and Wyoming. These studies will investigate brood production and migratory use of the ponds, and the results will be correlated with data on water quality, aquatic invertebrate populations, aquatic vegetation, and basin morphometry to help explain waterfowl use patterns. From these data, pond design and management criteria will be formulated.

Previous research (Evans and Kerbs 1977) suggests that construction of stock ponds on the northern Great Plains has greatly increased waterfowl habitat. Approximately 2 million puddle ducks—mallards, pintails, teal, and so forth—are produced annually on stock ponds constructed west of the Missouri River on the northern Great Plains.

Mine water impoundments can also be used by livestock. Those impoundments should be a minimum of one mile apart. Consequently, the spoils landform should be constructed to provide a water impoundment at least every mile. Trend surface analyses of the premined topography will suggest applicable spoils landforms that would provide the necessary topography and multiple use potential.

In summary, the overall objective of our research on the northern Great Plains is to assist land managers to: (1) successfully reestablish shrubs, trees, and forbs on mine spoils, and (2) design and manage water impoundments for a variety of beneficial uses.

NOTE

1. Research reported here was conducted in cooperation with: SEAM, EPA, ARCO, AMAX, Medicine Bow and Custer National Forest, South Dakota School of Mines and Technology, North Dakota State University, University of Wyoming, and several private landowners.

REFERENCES

Costello, David F. 1944. "Important Species of the Major Forage Types in Colorado and Wyoming," *Ecol. Manag.*, 14:107–134.

Evans, Keith E., and Roger R. Kerbs. 1977. *Avian Use of Livestock Watering Ponds in Western South Dakota.* Fort Collins, Colorado: General Technical Report, Rocky Mountain Forest and Range Experiment Station (in press).

Glass, G. B. 1972. "Midyear Review of Wyoming Coalfields," *Geol. Surv. Wyo. Annu. Rep.*

Jameson, D. A. 1952. "Nutritive Value of Browse on Montana Winter Ranges," *Journal of Range Management,* 5:306–310.

Landis, E. R. 1973. U.S. Senate Comm. on Inter. and Insular Affairs, "Coal in mineral and water resources of North Dakota." (93rd Congress, 1st Sess.)

Orr, Howard K. 1975. "Mine Spoil Reclamation at the Belle Ayre Mine, Northeast Wyoming," Proc. Fort Union Coal Field Symp., Vol. 3. Mont. Acad. Sci.

Packer, Paul E. 1974. *Rehabilitation Potentials and Limitations of Surface-Mined Land in the Northern Great Plains.* Ogden, Utah: Intermountain Forest and Range Experiment Station, USDA For. Serv. Gen. Tech. Rep. INT–14.

Thilenius, John F., and Gary B. Glass. 1974. "Surface Coal Mining in Wyoming: Needs for Research and Management," *Journal of Range Management,* 27:336–341.

Woolfolk, E. J. 1949. *Stocking Northern Great Plains Sheep Range for Sustained High Production.* U.S. Dep. Agric. Circ. 804.

Yamamoto, Teruo. 1975. "Coal Mine Spoil as a Growing Medium: AMAX Belle Ayre South Mine, Gillette, Wyoming," in Third Symp. on Surf. Min. and Reclam. Washington, D.C.: Vol. 1, NCA/BCR Coal Conf. and Expo II Natl. Coal Assoc.

2

Establishment of Native Plants for the Rehabilitation of Paraho Processed Oil Shale in an Arid Environment[1]

Cyrus M. McKell

Institute for Land Rehabilitation
Utah State University
Logan, Utah

INTRODUCTION

Oil shale deposits in Colorado, Utah, and Wyoming represent a great potential energy source that could be used by the United States to help meet increased energy requirements. In the winter and early spring of 1974, six 5,000-acre tracts of federal land in Utah, Colorado, and Wyoming were offered for prototype oil shale development leases in competitive bidding. Only the tracts in Colorado and Utah were leased for development. Tracts in Colorado were designated "Ca" and "Cb"; those in Utah, "Ua" and "Ub." Since 1974, considerable effort has been expended on environmental and engineering studies to guide the eventual development of these great energy resources.

Obviously, development of oil shale deposits has not yet reached the commercial stage at which batteries of retorts produce piles of processed shale to be rehabilitated or disturbed areas need to be revegetated. A baseline data-gathering program conducted on the oil shale tracts during 1975 and 1976 and predevelopment studies from 1975 to the present have, however, identified many problems that would be encountered should land rehabilitation be necessary. Thus, many techniques have been worked out in rehabilitation research that can be applied as oil shale development increases.

The purpose of this discussion is to describe: (1) some of the problems associated with rehabilitation of spent shale from the Paraho retort process under the arid conditions of the Uinta Basin of northeastern Utah, and (2) suitable techniques for establishing native shrubs on disposal piles and disturbed sites.

CRITERIA FOR REVEGETATION

Efforts to revegetate disturbed areas, disposal sites, construction projects, and even depleted rangelands in arid regions have generally met with failure because of the harsh environmental conditions and the lack of a suitable technology. Occasionally, some success has been achieved when an unexpected favorable turn of events, such as a wet year, has allowed seedling establishment. Even then, persistence of the seeded species has usually been short-lived because of incomplete plant adaptation, or unsuitable site preparation or maintenance.

Nevertheless, present societal attitudes and legislation require that lands disturbed in the process of development be rehabilitated. In the arid West, harsh sites and limited precipitation present a challenge to revegetation technology. In formulating any revegetation program, each site and its environmental constraints must be carefully considered.

The philosophy expressed in the National Environmental Policy Act (1969) declares a national commitment to the quality of the environment to be an objective concurrent with the regular missions of all federal agencies. In Utah, the objective of land rehabilitation is "to stabilize the land as quickly as possible after it has been disturbed in order to achieve permanent and protective vegetative cover. Nonnoxious native plants that will give a quick, permanent protective cover and enrich the soil shall be given priority." All state regulations call for some remedial action. The National Academy of Science Study Committee (1975) on the rehabilitation potential of Western coal lands states:

> . . . any decision to surface mine for coal in a particular area must include a strong commitment to rehabilitate the land concurrently. We concluded that there presently exists technology for rehabilitating certain western sites with a high probability of success, that others could be rehabilitated with lower probabilities of success, and that still others cannot be rehabilitated at all on the basis of what we know today. Into which category a particular site may fall depends critically on the site. . . .

A revegetation plan for rehabilitation of disturbed areas and disposal piles of processed oil shale must meet a number of criteria. First, it must fulfill the requirements outlined in the *Federal Register* (1973), which specify that the affected lands must be returned to usable, productive condition, which will be compatible with existing adjacent undisturbed natural areas and will support fauna of the same kinds and in the same numbers as those existing at the time the baseline data were obtained.

Second, the plan should make logical, effective use of natural resources available in each area, such as locally adapted plant species, precipitation, topsoil, and topography. A plan that either ignores the available resources (e.g., water or topsoil) or requires them in abundance may create new environmental impacts or require that present uses be restricted.

Third, a revegetation system should be cost-effective. By making optimum use of local resources and establishing a system that will be more or less self-sustaining after an initial development period, the costs of revegetation can be minimized.

THE ENVIRONMENTAL SETTING AND ITS LIMITATIONS

Vegetation existing on the oil shale tracts provides an index of the harshness of the environment there. Three main vegetation types—sagebrush/greasewood (*Artemisia/Sarcobatus*), juniper (*Juniperus*), and shadscale (*Atriplex*)—are typical in areas where drought, temperature extremes, and often salinity restrict plant establishment and growth. A fourth type, riparian, is highly productive because of the presence of water from the White River or Evacuation Creek. Any long-term revegetation technology for the Uinta Basin must consider the local environmental extremes and devise ways to overcome or work within their constraints.

Salt Desert Shrub, Sagebrush, and Juniper Vegetation

Several studies describe the types of vegetation on the tracts. Reports indicate a wide range of tolerance to environmental extremes exists among and within the various plant species that make up these three vegetation types, representative of about 55 percent of the plant cover of the arid West (Forest-Range Task Force 1972). Any generalizations about these types will be useful in making specific recommendations for revegetating disturbed areas and disposal sites on tracts Ua and Ub.

The salt desert shrub type is found on over 86 million acres of the saline valleys of the Great Basin and Colorado Plateau (Forest-Range Task Force 1972; Holmgren and Hutchings 1972). Annual precipitation averages 4–8 inches (100–200 millimeters [mm]), generally occurring as snow in the winter and infrequent thunderstorms in the summer. Temperatures are often below −37°C in the winter and above 40°C in the summer. Soils are sometimes deep but may also be shallow with minimal profile development. Soil salinity is ever-present and may range from slightly saline (4.0 milliohms per centimeter [mmhos/cm] electrical conductivity) to strongly saline (15.0 mmhos/cm electrical conductivity). Vegetative coverage, which amounts to 5–15 percent, is highly subject to seasonal variability because of abundant growth of annual species (weeds) in moist years. Productivity is limited and averages only 600 kilograms/hectare (kg/ha) on fair-condition sites (Mason 1971). In areas where salts accumulate, highly tolerant plant species occur, such as mat saltbush (*Atriplex corrugata*), iodine bush (*Allenrolfia occidentalis*), alkali suaeda (*Suaeda fruiticosa*), and winterfat (*Ceratoides lanata*). Where the plant community has been disturbed, invading annual species such as Russian thistle (*Salsola kali*), cheatgrass (*Bromus tectorum*), and halogeton (*Halogeton glomeratus*) have filled in the open spaces. Except for the poisonous halogeton, such weeds may be useful as temporary groundcover in the first stages of plant succession on rehabilitated areas.

Sagebrush/greasewood, as identified on tracts Ua and Ub, is not commonly found in the arid West as a single type. A big sagebrush (*Artemisia tridentata*)/bunchgrass type covers over 94 million acres of broad, level valley floors and uplands where deep soils of generally neutral pH value prevail (Forest-Range Task Force 1972). In areas where soils are shallow, rocky, and possibly more alkaline, other species of sagebrush, such as black sagebrush (*Artemisia nova*) occur. Greasewood, found in limited sagebrush areas, appears only on saline soils along dry stream courses or flats. Salinity in excess of 40 mmhos/cm has been measured in a greasewood area.

Annual precipitation ranges from 6 to 24 inches (150–600 mm) in areas where the sagebrush/greasewood type predominates. Winters are cold and summers are hot—the same temperature extremes as occur in the areas where the salt desert shrub type is found. Living plant cover percentage is low, about 15 percent, although not as low as in salt-desert-shrub-type areas. Productivity is low, averaging 850 kg/ha on fair-condition sites (Mason 1971). Root systems tend to be deep, although some species may have a shallow network.

The juniper vegetation type was thoroughly reviewed in a recent symposium at Utah State University (Gifford and Busby 1975). This type occurs as sparse-to-dense woodland on over 42 million acres (Forest-Range Task Force 1972) where precipitation averages 8–20 inches (200–500 mm). At the lower elevational extent of the type, few piñon pines (*Pinus* spp) are found because they are not adapted to the environmental extremes. Such is the case on tracts Ua and Ub. The arid upland soils on which junipers generally grow are shallow (60–90 cm) and are over fractured rocky parent material. The trees may also invade the deeper coarse soils of relatively level topography. Root systems tend to spread shallowly, competing with associated species for moisture. The soil reaction on most juniper sites is slightly alkaline.

Total vegetative coverage of a juniper type is low, often below 10 percent at the lower margin of its elevational distribution. Understory cover of forbs and grasses is less than 1 percent, and average production on fair-condition sites is 450 kg/ha (Mason 1971). Thus, a high proportion of soil surface in a juniper area is bare.

On tracts Ua and Ub, each of the three vegetation types discussed is adapted to similar environmental constraints: average annual precipitation of less than 10 inches (250 mm), cold winter and hot summer temperatures, and shallow saline or saline-alkaline soils. The adaptive capacities of these types also include: tolerance to grazing or browsing by animals; seed dispersal, germination, and establishment under conditions of a sparse vegetative canopy; and the ability to utilize occasional favorable years for extra increments of growth. (In spite of the last adaptive capability, the average productivity of the vegetation type, not the occasional high productivity of a good year, should serve as a guide to the expected productivity of a revegetated site.) These characteristics indicate that the native species of the area in the immediate vicinity of tracts Ua and Ub may represent the best choice of plants to be used for revegetation of disturbed sites and disposal areas.

Limited Resources

In addition to the harsh environmental conditions of the prototype oil shale lease areas, water is a major limitation to revegetation plans. In the arid West, water is vital for use in agriculture and industry, and for direct human use. A requirement for water in large quantities for irrigation or leaching of processed shale could only result in reallocation of water already committed to present uses. High priority must, therefore, be given to methods for efficiently using precipitation or reusing minor amounts of treated industrial process effluent.

Another limited resource of prototype lease land is topsoil. The soil survey of tracts Ua and Ub indicates the average soil depth is less than 18 inches (45 centimeters [cm]) on the upland areas. In the valley or canyon bottoms, soil material or alluvium is relatively deep (over 180 cm) but only the surface layer is biologically active. Even though a relatively large amount of coarse soil material may be present in the bottom of Southam Canyon—the area tentatively selected for disposition of processed shale— it would not be sufficient or available in a timely period for complete top-soiling of the disposal pile to the necessary depth. Inadequate topsoil was also noted in research reports of the Colony Development Operation by Bloch and Kilburn (1973). Stockpiling may be an alternative, but this practice could have undesirable effects on soil biological activities. In addition, costs of stockpiling and massive topsoiling are also important considerations that must be carefully examined. Any cost-reducing methods for use of topsoil will be valuable in overall cost projections.

Seeds of native species, both shrubs and understory plants, may be considered limited in that they are not commercially available. Seeds of native species and introduced weeds are present in the topsoil and would be available for natural revegetation wherever topsoil is used. Native seeds can also be collected to supplement seeds in the topsoil.

In summary, limited resources such as water, topsoil, and native seeds present certain constraints to the revegetation plan and the research required to validate the plan.

Problems Posed by Processed Oil Shale

Because of its physical and chemical properties, processed oil shale presents some potential problems to plant establishment. Data on TOSCO-processed shale indicate that it is highly saline with conductivities ranging from 9 to 26 mmhos/cm (Schmehl and McCaslin 1973; Bloch and Kilburn 1973). The texture and wetability of processed shale also depend on the processing method. TOSCO-processed shale is finer and more difficult to wet than processed shale from the U.S. Bureau of Mines gas combustion method or the somewhat similar Paraho process (Striffler, Wymore, and Berg 1974; Bloch and Kilburn 1973).

Processed shale is dark grey to black. Surface temperatures as high as 77°C (caused by the absorption of solar radiation) have been measured on test piles. Processed shale contains very little nitrogen or phosphorus but usually enough potassium for plant growth (Schmehl and McCaslin 1973). The sodium content of processed shale is high; this characteristic can not only affect plant growth but can be detrimental to the physical structure of the shale, causing dispersal of fine particles, decrease in permeability to water, and crusting of the shale as it dries. (Striffler, Wymore, and Berg 1974).

The Paraho process is one of the most probable types to be utilized on the prototype lease tracts in Utah. This process results in a dark grey to black, coarse, gravelly material, which crumbles easily. The larger particles break down physically through freezing-thawing and wetting-drying over a winter season. Some of the physical properties of the resulting processed shale were described in the detailed development plan presented by the White River Shale Project (1976). Some of the most striking differences between processed shale and typical topsoil from tracts Ua and Ub (Table 2.1) are: sodium and sulfate are relatively high; pH is as high as 9.0; and cation exchange capacity is low at 4.8 milliequivalents (Me)/100 g. The electrical conductivity of processed shale is also high—18 mmhos/cm.

Present Uses and Land Productivity

Present uses of the lease tracts consist of sheep and cattle grazing, wildlife habitat, recreation (hunting, touring/viewing), and watershed. Not all uses are prevalent to an equal extent, nor are they mutually exclusive. In terms of the major objective of returning the land to functioning

	Na^+	K^+	Ca^{++}	Mg^{++}	$SO_4^=$	Cl^-	HCO_3^-	$CO_3^=$	pH	B ppm
Processed oil shale	6.9	.6	1.4	1.9	11.1	.3	.2	0	9.0	.7
Topsoil	1.0	.1	.1	.3	.2	.1	.2	0	7.6	.9

Soluble ions (1:1 extract) me/100 g

	NO_3^- ppm	P ppm	K ppm	Fe ppm	Zn ppm	CEC me/100g	% N	Electrical Conductivity
Processed oil shale	.3	4.4	290	54	18.7	4.8	.18	18
Topsoil	130	21	297	6	7.8	11.3	.11	.7

Table 2.1   A Comparison of Chemical Parameters of Paraho Processed Oil Shale from Colorado and Topsoil from a Shadscale Plant Community on the Utah Oil Shale Prototype Lease Tracts Ua and Ub

productive condition, those uses existing at the time of the baseline study can be viewed as possible goals for the revegetation program.

Except in unusually wet years, such as 1976, the production of plant material is very low because of the harsh environment. Thus, revegetation of disturbed sites and disposal areas to a plant cover that fulfills the natural potential of the environment would be in conformity with the intention of the *Federal Register* requirements. The final result should be a vegetative cover that is capable of long-term stability without continued inputs of resources such as water and fertilizer and without the requirement of complete protection from animal use.

THE REVEGETATION CONCEPT AND UNDERLYING PRINCIPLES

A Revegetation Plan Must Be Scientifically Valid and Environmentally Compatible

The plan for Paraho processed oil shale revegetation reflects consideration of: concepts and practices in range management, watershed management, botany, and soil science; identified constraints found in the harsh, arid climate of the tracts; and the infertile, shallow soil that conditions the productivity of existing vegetation. The goal of the revegetation plan is to work within the environmental constraints of the ecosystems in establishing a vegetative cover that will (1) eventually be capable of perpetuating itself under natural conditions and (2) meet the needs of various users—as they existed at the time of the baseline study.

The Proposed System

The proposed system consists of seven compatible components that can function in a logical and coordinated fashion:

(1) *Use of the native plant species:* The best adapted species to meet the environmental extremes of the area are those that have survived to the present. Emphasis would be placed on using local ecotypes of dominant species having salinity tolerance, compatibility with domestic and wildlife use, and aesthetically desirable features.

(2) *Transplanting of container-grown stock:* The dominant species, usually shrubs and perennial grasses, would be grown from rooted cuttings clones, or seedlings. Plants with a root system of sufficient size to ensure survival would be transplanted at appropriate seasons of the year. Use of bare-root and/or container-grown stock avoids the high probability of failure in seed germination and seedling establishment involved in direct seeding in arid regions.

(3) *Surface shaping:* After compaction—to stabilize the pile and to reduce its permeability—the surface of the disposed shale pile would be shaped into terraces and slopes on side hills and into hillocks on the flat top (Fig. 2.1). This shaping would provide a catchment slope for harvesting precipitation and a terrace for transplanting plants.

(4) *Soil surface stabilization:* Mulches for binding surface particles, such as polyvinyl acetates, would be applied to catchment slopes and terrace edges. The surface treatment is expected to prevent movement of particles in the form of dust, assist or promote water harvest, and restrict the rise of salt from the process shale through capillary action.

(5) *Effective but limited use of topsoil:* A trench about 19 x 30 inches (45 x 75 cm) would be filled with topsoil. The necessary growing medium to receive the container-grown native plants, the topsoil would provide an inoculum of soil microorganisms, a partial source of native plant seeds, and a buffer against high salt concentrations that might develop from the processed shale. We would expect percolation of harvested water from the catchment slopes through the topsoil in the trench and out into the processed shale to keep salinity levels around the root zone within the tolerance of the transplanted native species.

(6) *Fulfillment of minimal fertility requirements:* Plants would be fertilized as needed and determined by existing and ongoing research.

CROSS SECTION

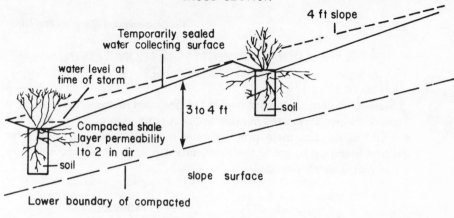

4 ft slope

Temporarily sealed
water collecting surface

water level at
time of storm

3 to 4 ft

soil

Compacted shale
layer permeability
1 to 2 in air

soil

slope surface

Lower boundary of compacted

CROSS SECTION

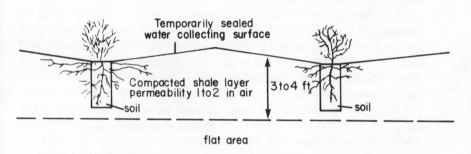

Temporarily sealed
water collecting surface

Compacted shale layer
permeability 1 to 2 in air

3 to 4 ft

soil

soil

flat area

Figure 2.1   Surface Modification and Vegetation on Processed Shale Slope Area and Flat
Area

Some fertilization would be necessary as plant roots grow into the
relatively infertile shale. (Because native plants of the region evolved in
relatively low fertility soils, their fertility requirements are not as high as
crop plants or some exotic species.)

(7) *Supplemental irrigation for plant survival and salinity management:*
As an added safeguard of plant survival during extended drought periods in
the early years of the revegetation project, supplemental irrigation may be
necessary. A drip-irrigation system, which would periodically supply low
volumes of water through a flexible pipe system installed along each
terrace (Hanks and Keller 1972; Bengson 1977), would be one way to
supply the water, if required. Depending on research findings, low-volume
irrigation may be needed to reduce salinity in the topsoil trench to levels
tolerable to the plants.

VALIDATION OF REVEGETATION CONCEPTS IN PUBLISHED REPORTS

Many of the concepts employed in the proposed revegetation plan have been developed and tested in other applications and in principle are valid for arid lands. The main problem now is to extrapolate criteria from existing literature to guide the revegetation program and suggest where additional research is needed.

Adaptation of Native Species to Extremes of Drought, Salinity, and Temperature

As stated, plant species native to salt desert shrub environments have evolved tolerances to extreme conditions; thus they are ideal for re-vegetating processed shale, where such tolerances may be vital for survival. Drought is accommodated partially by the development of an extensive root system, which may extract moisture from a large volume of near-surface soil or penetrate to deeper, moister soil layers (Kozlowski 1972). Moore, White, and Caldwell (1972) showed that two typical salt desert shrubs, *Atriplex confertifolia* and *Ceratoides lanata*, are able to extract water from soils having moisture potentials below −75 bars and to carry on photosynthesis when plant moisture potentials were below −115 bars. Fernandez (1974) observed root growth at soil moisture potentials as low as −70 bars in these desert shrub species. By way of comparison, most agronomic crop species wilt and cease most functions when soil moisture potential approaches −15 bars.

Salinity tolerance of salt desert shrub species is far greater than that of agronomic species. Growth may actually be stimulated by low salt concentrations, according to Wallace and Romney (1972) and Chatterton and McKell (1969). Maximum external salt concentrations tolerated by several salt desert species may approach or exceed the salinity level of seawater. Data shown in Figure 2.2. report the soil salinity levels where various plant species were found growing. Early reports of salinity were in percent salt of the dry soil or percent salt of the water extract. The salinity values of Figure 2.2. in mmhos/cm or parts per million were calculated according to methods developed by the U.S. Salinity Laboratory (Richards 1954). Although various salts may differ in their phytotoxicity, the conductivity measurements do not identify specific salts but only their combined osmotic strength. The salinity levels tolerated by salt desert shrubs indicate that they would be especially useful in vegetating the processed oil shale, which has been shown to have an electrical conductivity ranging from 9 to 26 mmhos/cm for its saturation extract.

The fertility of desert soils is inherently low, yet native shrubs flourish.

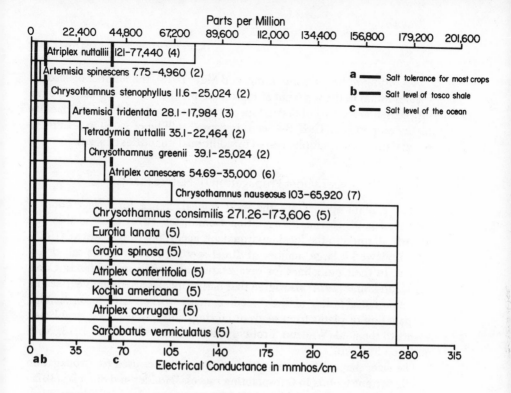

Figure 2.2 Maximum Reported Salt Tolerances of Selected Salt Desert Shrub Species. Sources: (2) Kearney et al. 1914; (3) Billings 1949; (4) Goodman 1973; (5) Shantz and Piemeisel 1940; (6) Wallace and Kleinkopf 1974; (7) Rollins, Dylla, and Eckert 1968

Goodman (1973) demonstrated that growth of salt desert shrubs is not increased by nitrogen fertilization and may actually be inhibited by the concomitant competition from weedy species such as *Halogeton glomeratus* and *Bromus tectorum*, which do respond to fertilization.

Wide temperature extremes are expected in the revegetation of processed shale. The small, reflective gray-green leaves of many salt desert species improve the ability of plants to dissipate heat and reduce the absorption of heat-producing solar radiation (Gates, Alderfer, and Taylor 1968). Leaf temperatures of over 50°C have been measured, but some species, such as *Atriplex confertifolia*, are still capable of active photosynthesis at these high temperatures (Caldwell 1972). Wein and West (1972) have shown that salt desert plants are able to contend with soil temperatures that sometimes exceed 60°C. Although seedlings could not survive such surface temperatures, transplants of container-grown shrubs are expected to do so and to modify gradually the conditions beneath their leafy cover to be suitable for establishment of seedling grasses and forbs.

According to Beauchamp, Lang, and May (1975), the number of seeds of species found in the top 5 cm of topsoil are plentiful for revegetation, but a supplemental seeding of desired species is necessary to obtain a satisfactory plant composition. Thus, the seeds in topsoil will be of value in leading to revegetation when supplemented by additional quantities of seeds.

Container Propagation of Native Plants

Native plant species are the most logical choice for revegetation of disturbed lands (Natural Vegetation Committee 1974) because they are already adapted to the local environment, according to Bleak et al. (1965), who reviewed a large number of direct seeding studies in the salt desert region. In their guidelines for revegetation of surface mined lands, Cook, Hyde, and Sims (1974) also agree that native species are desirable. Because of the potential failure of direct seeding, many workers have recommended propagation of plants from seeds or cuttings and transplanting established plants to the field (Wieland, Frolich, and Wallace 1971; Ellern 1973; Nord and Goodin 1970).

The size, shape, and composition of the container used for propagation are important factors in transplanting success. Hodder and Sindelar (1972) recommend a biodegradable tube in deep plantings for roadside stabilization in Montana. Containerized growth of plant materials for many types received considerable attention in a symposium organized by Tinus, Stein, and Palmer (1974). Unanimous agreement exists that transplanting container-grown plants is more successful than growing bare root stock, although many details must be worked out in order to satisfy the growth habits and other requirements of individual species. Planting of container-grown plants is the obvious choice for revegetating harsh sites.

Water Harvesting and Runoff

Creating slopes, hillocks, and irregular surface configurations is an accepted practice in landscape design (McHarg 1971), which gives aesthetic appeal and provides a more natural appearance to disposal areas. Far more important in an arid region is the need to contour and slope the surface to harvest precipitation for use by plants. Such a practice was well developed as "runoff farming" in the Negev Desert by various peoples as early as 2000 B.C. (Evenari, Shanon, and Tadmoor 1971). Of particular interest is the concept of microcatchments and the results obtained by Evenari and his associates. They found that in an area where the average precipitation is 3.3 inches (83 mm), *Atriplex halimus* transplants survived and produced 400 kg/ha of new growth (dry weight basis) each year. The optimum catchment size was 31.2 square meters (m^2).

Recently Aldon and Springfield (1975) have shown excellent survival rates for transplanted species in basin catchments surface-treated with paraffin or polyethylene. Adaptation of the water-harvesting concept to the rehabilitation of processed oil shale piles will require local studies on catchment slope, size, and location of plants in relation to salt movement.

Surface Stabilization with Mulches

Using mulches of various kinds has become a routine practice to hold topsoil in place, provide a seedbed, or prevent dust. Wood fibers, waste papers, excelsior mats, and so forth are used to stabilize slopes, often with tackifiers, for landscaping or roadside stabilization projects (Hodder and Sindelar 1972; Kay 1973, 1974).

Of particular relevance to the stabilization of processed oil shale piles is the possibility of using a surface mulch material that not only would stabilize small particles that are sources of dust but would serve to harvest precipitation. Reports by U.S. Bureau of Mines workers (Havens and Dean 1969; Dean, Havens, and Glantz 1974), Kay (1975), and Frasier and Myers (1970) indicate that some materials are good surface stabilizers. Whether their sealing capacities are sufficient to harvest water also requires further research. A recent symposium on water harvesting provides information on many materials and methods applicable to the present problem with oil shale (Frasier 1974).

Growth of Plants in Processed Shale

Because of the harshness of the medium, most studies of plant growth in spent shale have dealt with extensive modifications of the shale—such as leaching salts with large amounts of water, heavy fertilization, topsoiling, and mulching (Harpert and Berg 1974; Bloch and Kilburn 1973). Through the use of these treatments, often supplemented by periodic irrigation, a large number and variety of plant species have been successfully grown on processed shale. The cost of such treatments would be high, however (Heley and Kilburn 1973), considering the large deposits of processed shale expected from commercial oil shale production.

In one of the few studies using unleached shale, Schmehl and McCaslin (1973) observed that seed germination of tall wheatgrass (*Agropyron elongatum*) and Russian wildrye (*Elymus junceus*) in the greenhouse was 18 percent and 29 percent, respectively, on pure processed shale as compared to 100 percent on soil. The subsequent growth of both species on processed shale was only 1–2 percent of that on soil. Merkel (1973) also reported poor establishment of vegetation by direct seeding on unleached and unirrigated spent shale in the field, although the plant growth on soil was good. Interestingly, however, Baker and Duffield (1973) noted the successful invasion of several weedy species on nonirrigated and nonfertilized spent shale plots, indicating that adapted species could be established with only minimal treatment of the spent shale. Further research to determine the minimum treatments necessary for successful establishment and maintenance of the acceptable vegetation on spent shale is needed.

Ecological Succession of Rehabilitated Areas

Implied in the requirements for revegetation of disturbed sites and disposal areas is the development of a stable, productive plant cover. Considerable time is needed to assure the ecological stability of a reestablished plant community. Follow-up studies of seeded rangelands in the West have indicated that some of the seeded species did not survive. Annual weedy species may persist and compete strongly with native and seeded species, and the return to a stable condition may require a long time (Bleak et al. 1965; Hull 1973; Robertson and Pearse 1945). Although undesirable from some points of view, the invasion of annual species may modify some extreme factor of the environment of a disturbed site and thus render it more favorable for succeeding species. Such is the philosophy of initial plant succession (Weaver and Clements 1938).

RESULTS OF PRESENT RESEARCH AT USU ON REVEGETATION OF PROCESSED SHALE

Research at USU on problems related to oil shale revegetation has been under way for only a few years. This work draws, however, upon a large number of studies of arid (range) land management, ecology, and revegetation over a period of many years. The present work provides additional answers to problems in five areas: (1) identification of some desirable native species for revegetation, (2) propagation methods, (3) planting techniques, (4) plant growth characteristics, and (5) soil surface coatings for stabilization. Some applications of this research to the revegetation program are described in the following sections.

Identification of Desirable Species

Adaptation to existing and predicted environmental extremes is a prerequisite for species to be used in revegetation. Natural selection has already provided a wide array of shrubs, grasses, and forbs from which to choose. Selection criteria have been applied to identify dominant local species that: are tolerant to salinity, have desired rooting habits, are desirable feed for wildlife and/or livestock and have tolerance to utilization, possess a degree of natural beauty, can be propagated easily, may be established successfully, and are compatible with other species.

Studies have identified local dominant species and the proportion of the floral composition they constitute. Field data may serve as a guide to the selection of species and their relative proportions in the vegetative cover to be established.

Some of the species presently identified for use include:

GRASSES

Agropyron smithii
Bromus tectorum
Distichlis stricta
Elymus junceus
Elymus salina
Hilaria jamesii
Oryzopsis hymenoides
Sporobolus cryptandrus

FORBS

Atriplex spp
Bassia hyssopifolia
Hedysarum boreal
Kochia spp
Salsola kali
Sphaeralcea spp
Suaeda fruiticosa

SHRUBS

Artemisia frigida
Artemisia nova
Artemisia tridentata
Atriplex canescens
Atriplex confertifolia
Atriplex corrugata
Atriplex cuneata
Chrysothamnus nauseosus
Chrysothamnus greenii
Chrysothamnus visicidiflorus
Ceratoides lanata
Grayia brandegei
Glossopetalon nevadensis
Sarcobatus vermiculatus
Juniperus osteosperma

Plant Propagation

Native species may be propagated from rooted cuttings or germinated seeds to provide transplanting stock of desired species. Because transplanted plants survive better than those directly seeded, methods for propagating and growing of local native species are needed.

Most shrub species studied require higher concentrations of rooting hormone (1–2 percent) than horticultural species. A sterilized compressed peat pellet (Jiffy #7) is an excellent rooting medium, which can later be transferred with the rooted cutting into a container. Efforts to root two important shrubs, *Artemisia tridentata* and *Sarcobatus vermiculatus,* have produced poor results so far. Other species studied appear to root moderately well to excellently.

Seed germination of species listed above is relatively effective using techniques generally outlined in the *Handbook for Seeds of Woody Plants*

(Schopmeyer 1974). *Atriplex* seeds are often a problem, however, because of an indurate utricle (a hard outer seed layer); seed fill is also low and consequently a large volume of seeds must be processed to yield a smaller number of quality seeds. This problem is presently under intensive study to refine techniques for discarding unfilled seeds and those of low quality.

Harvesting and cleaning methods have been devised for the seeds of dominant native plants of the tracts.

Planting Techniques

As stated, transplanting of container-grown plants gives better results than either direct seeding or bare-root seedling transplants. A wide choice of planting times during late spring and early summer is possible when container-grown transplants with roots intact in a soil mass are placed in 8-inch-deep holes and irrigated with one liter of water. Survival rates of nearly 100 percent have been obtained on disturbed sites. Paraho processed shale for field transplanting research has not been available for extensive planting.

Plant Growth Characteristics

Preliminary results and observations of plant growth habits provide useful information to guide selection and propagation studies, transplanting methodologies, and design of the system for shaping the surface of the shale disposal pile. Root systems of the major shrub species have been observed and classified as to depth and spread.

Some of the native species of the tract will grow in processed shale according to preliminary greenhouse tests. *Atriplex* species appear to be the best. Russian wildrye is a good performer among the grass species tested.

Bromus tectorum, an introduced annual weedy grass prevalent in the tract areas, will grow well on a 1:1 mix of processed shale and soil. Growth is about 90 percent less on unleached shale with no fertilizer. Adding nitrogen and phosphorus will somewhat increase growth. *Bromus tectorum* may be useful as a colonizer species.

Soil Surface Stabilizing Materials

Materials to coat the soil surface and thereby stabilize dust-forming particles and give a degree of impermeability to the surface have been studied under field conditions. Two polyvinyl acetate formulations have proved useful. After four months, the materials were still effective in particle stabilization and to a lesser degree in watershedding. After a season of overwintering, the materials lost most of their effectiveness. Refinements in stabilization rates, methods of application, and comparison of related materials are needed.

SUMMARY

Rehabilitation methods for disturbed areas and disposal piles of processed oil shale have been devised in advance of the development of extensive oil shale deposits in Colorado, Utah, and Wyoming. These methods must necessarily be tentative until actual development takes place that would provide conditions for implementation.

A methodology emphasizing native plants and minimal use of expensive practices is proposed based on principles from various scientific fields and ongoing research. The basic components of the rehabilitation plan discussed in this paper include the following:

(1) Use of native plant species that are adapted to the harsh environment of the oil shale region

(2) Transplanting of container-grown plants to avoid the low rate of success often experienced with direct seeding

(3) Soil surface shaping to create small slopes for surface runoff and terraces for infiltration and plant establishment

(4) Soil surface stabilization to promote runoff from small slopes and to reduce polluting soil particle movement

(5) Use of topsoil on a limited basis in trenches where plants can be grown rather than a thin overall application to disposal areas

(6) Application of fertilizers only in sufficient quantities to meet the nutritional requirements of native plants

(7) Supplemental irrigation as needed to insure plant establishment in drought conditions.

Modification of these components may be necessary under actual operating conditions of oil shale development.

NOTE

1. Acknowledgments: the author gratefully acknowledges the assistance of colleagues in preparing this paper: Gordon Van Epps, Associate Professor of Range Science; Kent Crofts, Research Technician; Steven Richardson, Research Technician; and Jerry Barker, Graduate Research Assistant.

Concepts and data reported in this paper are from a project partially funded by the White River Shale Project in a joint venture with Sun Energy Development Company, SOHIO Petroleum Company, and Phillips Petroleum Company.

REFERENCES

Aldon, E. F., and H. W. Springfield. 1975. "Using Parafin and Polyethylene to Harvest Water for Growing Shrubs," in Proceedings of Water Harvesting Symposium, Phoenix, Arizona: USDA-ARS W-22.

Baker, F., and W. J. Duffield. 1973. "Annual Revegetation Report," in Bloch, M. B., and P. D. Kilburn [eds]. *Processed Shale Revegetation Studies, 1965–1973*. Denver: Colony Development Operation, Atlantic Richfield Co.

Beauchamp, H., R. Lang, and M. May. 1975. *Topsoil as a Seed Source for Reseeding Strip Mine Spoils*. Wyoming Agric. Exp. Sta. Res., Journ. 90.

Bengson, Stuart A. 1977. "Drip Irrigation to Revegetate Mine Wastes in an Arid Environment." *Journal of Range Management*, 30:143–7.

Billings, W. D. 1949. "The Shadscale Vegetation Zone of Nevada and Eastern California in Relation to Climate and Soil, *Amer. Midland Nat*, 42:87–109.

Bleak, A. T., N. C. Frischknecht, A. P. Plummer, and R. E. Eckert. 1965. "Problems in Artificial and Natural Revegetation of the Arid Shadscale Vegetation Zone of Utah and Nevada, *Journal of Range Management*, 18:59–65.

Bloch, M. B., and P. D. Kilburn [eds]. 1973. *Processed Shale Revegetation Studies, 1965–1973*. Denver: Colony Development Operation, Atlantic Richfield Co.

Caldwell, M. W. 1972. "Gas Exchange in Shrubs," in McKell, C. M., J. P. Blaisdell, and J. R. Goodin [eds], *Wildland Shrubs—Their Biology and Utilization*. USDA Forest Service Gen. Tech. Rep. INT-1.

Chatterton, N. J., and C. M. McKell. 1969. "*Atriplex Polycarpa*. I. Germination and Growth as Affected by Sodium Chloride in Water Cultures." *Agron. Journal*, 61:448–450.

Cook, C. W., R. M. Hyde, and P. L. Sims. 1974. *Guidelines for Revegetation and Stabilization of Surface Mined Lands in the Western States*. Fort Collins, Colorado: Range Sci. Ser. No. 16. Colorado State University.

Dean, K. C., R. Havens, and M. W. Glantz. 1974. *Methods and Costs for Stabilizing Fine-Sized Mineral Wastes*. U.S. Bureau Mines Rep. of Invest. 7896.

Ellern, S. J. 1973. "Rooting cuttings of saltbush (*Atriplex halimus* L.)," *Journal of Range Management*, 25:154–155.

Evenari, M., L. Shanan, and H. Tadmoor. 1971. The Negev—the Challenge of a Desert. Cambridge, Massachusetts: Harvard University Press.

Federal Register, 1973. 38:230, Part III, Section II. Washington, D.C.: U.S. Government Printing Office.

Fernandez, O. A. 1974. *The Dynamics of Root Growth and the Partitioning of Photosynthesis in Coal Desert Shrubs*. Ph.D. Dissertation. Logan, Utah: Utah State University.

Forest-Range Task Force. 1972. *The Nation's Range Resources—A Forest Range Environmental Study*. Washington, D.C.: Forest Resources Rep. No. 19, USDA Forest Service.

Frasier, G. W. [ed]. 1974. *Proceedings of the Water Harvesting Symposium*. Phoenix, Arizona: USDA-ARS W-22.

Frasier, G. W., and L. E. Myers. 1970. *Protective Spray Coatings for Water Harvesting Catchments*. Trans. Am. Soc. Agr. Eng. 13:292–294.

Gates, D. M., R. Alderfer, and E. Taylor. 1968. "Leaf Temperatures of Desert Plants," *Science*, 159:994–995.

Gifford, G. F., and F. E. Busby. 1975. *The Pinyon-Juniper Ecosystem: A Symposium*. Logan, Utah: Utah State University, College of Natural Resources and Agric. Exp. Sta.

Goodman, P. J. 1973. "Physiological and Ecotypic Adaptations of Plants to Salt Desert Conditions in Utah," *J. Ecology*, 61:473–494.

Hanks, R. J., and J. Keller. 1972. "New Irrigation Method Saves Water but It's Expensive," *Utah Sci.*, 33:79–82.

Harpert, H. P., and W. A. Berg. 1974. *Vegetative Stabilization of Spent Oil Shales.* Fort Collins, Colorado: Env. Resources Center Tech. Rep. No. 4. Colorado State University.

Havens, R. and K. C. Dean. 1969. *Chemical Stabilization of the Uranium Tailings at Tuba City, Arizona.* U.S. Bureau of Mines Rep. of Invest. No. 7288.

Heley, W. and P. D. Kilburn. 1973. "The Cost of Processed Shale Revegetation," in Bloch, M. B., and P. D. Kilburn [ed]. *Processed shale revegetation studies, 1965–1973.*

Hodder, R. L., and B. W. Sindelar. 1972. *Tubelings—A New Dryland Technique for Roadside Stabilization and Beautification.* Montana Agric. Exp. Sta. Res. Rep. No. 18. Bozeman, Montana: Montana State University.

Holmgren, R., and S. S. Hutchings. 1972. "Salt Desert Shrub Response to Grazing Use," in McKell, C. M., J. P. Blaisdell, and J. R. Goodin [eds]. *Wildland Shrubs—Their Biology and Utilization.* USDA Forest Service Gen. Tech. Rep. INT-1. 153–164.

Hull, A. C. 1973. "Duration of Seeded Stands on Terraced Mountain Lands, Davis County, Utah," *Journal of Range Management,* 26:133–136.

Kay, B. L. 1973. *Wood Fiber Mulch Studies.* Davis, California: Agronomy Prog. Rep. No. 52. California Agric. Exp. Sta.

Kay, B. L. 1974. Davis, California: *Erosion Control Treatments on Coarse Decomposed Granite.* Agronomy Prog. Rep. No. 60. California Agric. Exp. Sta.

Kay, B. L. 1975. *Tackifiers for Straw Mulch.* Davis, California: Agronomy Prog. Rep. No. 63. California Agric. Exp. Sta.

Kearney, T. H., L. J. Briggs, H. L. Shantz, J. W. McLane, and R. L. Piemeisel. 1914. "Indicator Significant of Vegetation on Tooele Valley, Utah," *J. Agr. Res.,* 1:365–417.

Kozlowski, T. T. 1972. "Physiology of Water Stress," in McKell, C. M., J. P. Blaisdell, and J. R. Goodin [eds]. *Wildland Shrubs—Their Biology and Utilization.*

Mason, L. 1971. *Yield and Composition of Utah's Range Sites.* Soil Conserv. Serv. Mimeo Rep.

McHarg, I. 1971. *Design with Nature.* Doubleday/Natural History Press.

Merkel, D. L. 1973. "Performance of Plants with Minimal Treatment of Processed Oil Shale: An Interim Report," in Bloch, M. B., and P. D. Kilburn [eds]. *Processed Shale Revegetation Studies, 1965–1973.*

Moore, R. M., R. S. White, and M. M. Caldwell. 1972. "Transpiration of *Atriplex confertifolia* and *Eurotia lanata* in Relation to Soil, Plant, and Atmospheric Moisture Stresses," *Can. J. Bot.* 50:2411–2418.

National Academy of Sciences Study Committee. 1974. *Rehabilitation Potential of Western Coal Lands.* Cambridge, Mass.: Ballinger Publ. Co.

National Environmental Policy Act of 1969. P.L. 91–190.

Natural Vegetation Committee, Arizona Chapter, SCS. 1973. *Landscaping with Native Arizona Plants.* Tucson, Arizona: Univ. of Arizona Press.

Nord, E. C., and J. R. Goodin. 1970. *Rooting Cuttings of Shrub Species for Plantings in California Wildlands.* Berkeley, California: USDA For. Serv. Res. Note PSW 213.

Richards, L. A. [ed] 1954. *Diagnosis and Improvement of Saline and Alkali Soils.* USDA Agric. Handb. No. 60.

Robertson, J. H., and C. K. Pearse. 1945. "Range Reseeding and the Closed Community," *Northwest Sci.,* 19:58–66.

Rollins, M. B., A. S. Dylla, and R. E. Eckert. 1968. "Soil Problems in Reseeding a Greasewood-Rabbitbrush Range Site," *Journal of Soil and Water Conservation,* 23(4):138–140.

Schmehl, W. R., and B. D. McCaslin. 1973. "Some Properties of Spent Oil Shales Significant to Plant Growth," in Hutnik, R. J., G. Davis [eds]. *Ecology and Reclamation of Devastated Land,* Vol. 1. New York: Gordon and Breach.

Schopmeyer, C. S. 1974. *Seeds of Woody Plants in the United States.* USDA Forest Service Agric. Handbook No. 450.

Shantz, H. L., and R. L. Piemeisel. 1940. *Types of Vegetation in Escalante Valley, Utah, as Indicators of Soil Conditions.* USDA Tech. Bull. 1973.

Striffler, W. D., I. F. Wymore, and W. A. Berg. 1974. "Characteristics of Spent Shale Which Influence Water Quality, Sedimentation, and Plant Growth Medium," in Cook, C. W. [ed]. *Surface Rehabilitation of Land Disturbances Resulting from Oil Shale Development.* Fort Collins, Colorado: Env. Res. Center Tech. Ser. No. 1.

Tinus, R. W., W. I. Stein, and W. E. Palmer. 1974. *Proceedings of the North American Containerized Forest Tree Seedling Symposium.* Denver, Colorado, August 26-29. Great Plains Agricultural Council Publication, No. 68.

Wallace, A., and E. M. Romney. 1972. *Radioecology and Ecophysiology of Desert Plants at the Nevada Test Site.* Oak Ridge, Tennessee: U.S. Atomic Energy Commission.

Wallace, A., and G. E. Kleinkopf. 1974. "Contributions of Salts to the Water Potential of Woody Plants," *Plant Sci. Letters,* 3:251–257.

Weaver, J. E., and F. E. Clements. 1938. Plant Ecology. 2nd Ed. New York: McGraw-Hill Book Co.

Wein, R. W., and N. West. 1972. "Physical Microclimates of Erosion Control Structures in a Salt Desert Area," *J. Appl. Ecol.* 9:703–719.

White River Shale Project. 1976. *Detailed Development Plan: Federal Lease Tracts Ua and Ub.*

Wieland, P. A. T., E. F. Frolich, and A. Wallace. 1971. "Vegetation Propagation of Woody Shrub Species from the Northern Mojave and Southern Great Basin Deserts," *Madrono* 21(3):149–152.

3

Restoration of Productivity to Disturbed Land in the Northern Great Plains

J. F. Power, Fred M. Sandoval, and Ronald E. Ries

Northern Great Plains Research Center
USDA-ARS
Mandan, North Dakota

SUMMARY

This paper presents a review of current information on the potential for restoring agricultural productivity to land disturbed by strip mining in the Northern Great Plains. The discussion focuses on the nature of the available natural resources, and information is presented to show how these natural resources can be recombined to return the land to productive uses.

INTRODUCTION

The use of fossil energy is rapidly increasing throughout the world. In the United States, this increase has averaged about 4 percent annually over the past several decades. During this period, much of the rising demand has been met by increased use of petroleum products. However, U.S. crude oil reserves are rapidly being depleted, necessitating the importation of foreign oil. Imported oil will soon account for 50 percent of the oil consumed in the United States. Consequently, there is a growing recognition of the need to maintain a favorable balance of foreign trade, to lessen dependence of the U.S. economy on actions by foreign governments, and to conserve the remaining domestic oil supplies.

Recently, considerable interest and economic stimulation have been given to the substitution of coal for oil. Known U.S. coal reserves are sufficient to meet anticipated energy requirements for several centuries (Averitt 1975). Many of these reserves are located in the Western United States, especially in the Northern Great Plains. These coal fields, containing lignitic and subbituminous coal, have remained essentially undeveloped until the last decade.

The development of coal fields in the Northern Great Plains is not, however, without social, economic, and environmental problems. The area is sparsely populated (Table 3.1), and has an adequate but widely dispersed transportation system. The region is dependent primarily on an agricultural economy and lacks an industrial labor market. However, the coal deposits are vast (Table 3.2), with coal seams commonly 1 to 10 meters (m) thick—and some exceeding 30 m. The deposits are located in a landscape well suited to surface mining. This coal is relatively low in sulfur (S) content, reducing or eliminating the need for sulfur dioxide (SO_2) scrubbers when the coal is burned. The cost per calorie of burning surface-mined Western coal is frequently 30–50 percent less than that of Eastern U.S.–mined coal (National Coal Association 1975). Thus, these coal fields are rapidly being developed, with the coal either burned in large electrical-power-generating plants near the mine sites or transported by 10,000-ton unit trains to industrial centers in the central states. Also, permits have been granted to construct the first of potentially many coal gasification plants that will convert coal to a pipeline-quality gas (methane).

These factors are resulting in the development of extremely large mining operations. Most mines presently operating in the Northern Great Plains produce near or over 1 million tons of coal annually, and those

Table 3.1   Area and Population of Coal-Producing States in the Western United States (Source: 1970 census)

State	Area	Population	Population density
	thousand km^2	thousand	no./km^2
North Dakota	183	618	3.4
Montana	381	694	1.8
Wyoming	254	332	1.4
Colorado	270	2,207	8.2
Utah	220	1,059	4.8
New Mexico	315	1,016	3.2

		Recoverable	
State	Total estimated	Surface Mines	Underground	Total
		- - - - - - - - -million tons - - - - - - - - - - - - - -		
North Dakota	560,630	16,003	0	16,003
Montana	378,675	42,562	65,165	107,727
Wyoming	545,656	23,674	27,554	51,228
Colorado	371,659	870	14,000	14,870
Utah	79,721	262	3,780	4,042
New Mexico	109,427	2,258	2,136	4,394

Table 3.2   Coal Reserves in Several of the Western United States (Source: National Coal Association 1975)

presently being developed will produce 5–20 million tons annually. Because of the scale of these operations and the economics involved, almost all mining is now done using surface-mining techniques. Collectively, these factors have resulted in rapid acceleration in coal production in the Northern Plains (Table 3.3), and an increase in the area disturbed annually by surface mining.

Most land in the Northern Great Plains is used primarily for the production of domestic livestock (cattle and sheep) or is cultivated for dryland cropping, especially wheat and barley. Secondary uses include wildlife habitat and recreation. Because it is likely that the same land uses will be desired after mining, the rapid growth of mining is accompanied by the need to develop technology to restore to the land permanently the capacity to produce useful vegetation for the benefit of present and future generations. The harvest of coal is a one-time operation, while the harvest of agricultural produce and wildlife from this same land is an annual event.

Historically, surface mining for coal in the Northern Great Plains began 50 to 70 years ago at scattered locations to provide fuel for locomotives and for domestic consumption by pioneer homesteaders. Because only a few hundred hectares were mined annually, and because a frontier of new land existed, there was little concern for reclamation. Since the development of Western coal for electrical power generation and other industrial uses within the last decade, however, thousands of hectares are disturbed annually. This area will soon increase to tens of thousands of hectares. At the same time, concern over food production and awareness of aesthetic values and of the need for environmental protection have increased.

Production in:

State	1962	1966	1970	1973	1980[1]	2000[1]
			-thousand	tons		
North Dakota	2,733	3,543	5,639	7,400	19,000	119,000
Montana	382	419	3,447	9,950	41,000	133,000
Wyoming	2,569	3,670	7,222	13,600	47,000	110,000

[1] At projected intermediate rate of development (Northern Great Plains Resource Program, 1974).

Table 3.3   Recent and Projected Annual Coal Production in the Northern Great Plains

Consequently, most Northern Great Plains states involved with mining have recently enacted reclamation laws, with the intent of restoring the landscape to a condition of equal or greater potential productivity for agriculturally important plant species than existed before mining.

Because of the recency of the awareness of the need for reclamation, present reclamation technology is actually in its infancy. Extensive research in most major mining regions has defined the primary problems involved and has suggested potential solutions for these problems. New practices that require a change in mining methods or equipment are sometimes slow and difficult to implement, however, because of the huge scale of major mining operations. Frequently, at these mines, over 30 million tons of earth and coal are moved annually. Huge, costly equipment is required for the operations. Fabrication and replacement of equipment to enable new practices usually requires several years and is extremely expensive.

RESOURCES

Most major coal mining regions in the Northern Great Plains have a continental climate. Table 3.4 summarizes several parameters of climate at several recording stations located in or near some major Western coal fields. Generally, annual precipitation is in the 20-to-40-centimeter (cm) range, with highest rainfall occurring between April and August. Most coal fields are located either in northern latitudes or at elevations greater than 1,000 m—areas with relatively short growing seasons, cold winters, and often hot summers.

Perennial grasses are adapted to almost all regions, but because of

Location	Annual precipitation	Month highest precipitation	Mean July temperature	Frost-free season 0°C
	mm		°C	days
Dunn Center, ND	399	June	21.0	115
Circle, MT	287	June	21.2	99
Colstrip, MT	384	June	22.6	127
Gillette, WY	363	June	22.4	129
Evanston, WY	269	May	17.1	49
Hayden, CO	391	April	18.1	76
Escalante, UT	310	August	21.4	138
Bloomfield, NM	208	August	23.3	147

Table 3.4   Mean Values of Selected Parameters of Climate near Several Major Coal Fields in the Western United States

climatic limitations, production of cultivated crops is limited to the areas with more favorable water supplies. Some shrubs and forbs are also adapted to the regions, while trees are confined to sandy areas or protected sites. Various wheatgrass species (*Agropyron*) are adapted throughout these areas, with crested wheatgrass (*A. desertorum* [Fisch] Schult.) being the species most commonly seeded. Other wheatgrass species found include slender (*A. trachycaulum*), western (*A. smithii* [Aybd.]), and intermediate (*A. intermedium*) wheatgrass. Grama species (*Bouteloua gracilis* [H.B.K.] Lag. and *B. curtipendula* [Michx] Torr.) are frequently found in native vegetation. In addition to many of the above, species of *Stipa, Poa, Bromus, Panicum, Sporobolus,* and other genera are often seeded on reclaimed mined lands. Legumes, such as alfalfa (*Medicago sativa* L.) and sweetclover (*Melitotus officinalis* L.), are frequently grown. Annual crops commonly grown include hard red wheats (*Triticum aestivum* L.) and barley (*Hordeum vulgare* L.).

In addition to climate and adapted vegetation, another basic resource involved in reclamation of mined land is the soil and overburden. Sandoval et al. (1973) and others provided information on various physical and chemical properties of coal mine spoils and overburden in the Northern Great Plains. Typical data on characteristics of spoils at several mine sites are given in Table 3.5. It is evident that many characteristics are site-specific, varying from mine to mine—or within a mine. However, most overburden and spoils in the Northern Great Plains have certain character-

Location	pH	CaCO₃ Equiv-alent	Clay	EC	Saturation Extract:				SAR [1]
					Ca	Mg	Na	SO₄	
		%	%	mmhos/cm²	------	------	meq/l	------	------
Beulah, ND	8.4	12	37	3	< 1	< 1	34	30	41
Center, ND	7.6	10	26	4	14	29	7	48	2
Stanton, ND	8.3	12	52	2	1	1	20	16	19
Zap, ND	8.8	10	54	2	< 1	< 1	19	7	48
Colstrip, MT	7.2	-	-	3	15	19	3	-	1
Sheridan, WY	6.1	-	-	7	29	28	45	-	8
Gillette, WY	7.4	-	-	8	25	73	31	-	5

[1] SAR = Sodium adsorption ratio = $Na/\sqrt{Ca + Mg/2}$, and is highly correlated with exchangeable sodium percentage (ESP).

Table 3.5 Properties of Surface Mine Spoils at Several Northern Great Plains Mine Sites (Source: Sandoval et al. 1973)

istics in common—they are neutral or alkaline in reaction (pH), contain appreciable calcium (Ca) salts (especially calcium carbonate [$CaCO_3$]), and contain variable quantities of soluble salts (mainly comprising varying ratios of Ca, magnesium [Mg], and sodium [Na] sulfates). They are almost universally deficient in plant-available phosphorus (P) and biologically active forms of organic nitrogen (N), but may contain appreciable amounts of inorganic N (Power et al. 1974). Texture (particle size distribution) varies widely from location to location, ranging from clays to sands. Pyrite-bearing minerals and subsequent problems of acidity are almost entirely absent from these spoils and overburden. Most of these coal fields are located in the Fort Union geologic group.

The nature of the soil resources in the Northern Great Plains also varies widely, and includes various borolls and aridisols, as well as badlands. Natric (sodic) and saline soils are not uncommon in the region. Parent materials are typically the sandstones, siltstones, and shales that characterize the Fort Union geologic group. However, in localized areas, this formation may be covered with more recent alluvium, glacial drift, loess, or other materials.

With knowledge of plant growth requirements, a logical approach to the reclamation of strip-mined land is to: (1) inventory, sample, and map the original soils, and determine the physical and chemical properties of the overburden before mining; (2) relate the properties of the original soils and overburden, and their distribution over the area, to potential productivity of plant species that are of major significance when grown under the prevailing climate; and (3) devise a mining plan that will economically extract the coal resources and leave the landscape in a condition that permits the later development of the full productivity potential of the disturbed area. The decision of whether or not to develop fully this productivity potential can be left to the landowner, just as he now makes this decision on unmined land. Potentially irrigable land, for example, would be restored to a condition that would permit this potential use. However, present and future landowners would have the option of deciding whether or when to make the investment needed to develop irrigated agriculture. This approach provides the maximum potential for utilization of the available natural resources, and allows present and future generations the flexibility to make decisions in regard to land use.

WATER

Throughout the arid and semiarid regions of the Western United States, reclamation of land disturbed by strip mining depends primarily on the conservation and efficient utilization of the limited precipitation received.

On unmined land in the Northern Great Plains, it is usually lack of water that ultimately limits plant growth. Practices and techniques designed for the proper conservation and efficient use of water by productive vegetation are needed to restore adequate productivity to mined land. Reclamation technology must: enhance infiltration; reduce runoff, soil water evaporation, and leaching; and increase root absorption of soil water and dry matter production per unit of water used.

Water Storage

The storage of water received as precipitation is controlled by those properties of mine spoils that affect infiltration, runoff, erosion, and sedimentation. Included are texture, extent of fragmentation of sedimentary rocks in the overburden, exchangeable Na content, bulk density, and slope and slope length. As mentioned, texture of spoils varies widely from mine to mine, and even within a mine. Infiltration generally decreases as clay content increases, but closely packed fine sand fractions, even with relatively low clay contents, can also be relatively impermeable. Most spoils are derived from sedimentary rock—sandstones, siltstones, and shales. However, the degree of fragmentation and fracturing of these materials during the mining process may affect infiltration. Shales may fracture into pieces ranging from a pulverized, almost single-grain arrangement to segments several centimeters in diameter, depending on their properties and on the mining technique. Consequently, differential packing of these fragments results in variable porosity and infiltration rates. The larger fragments frequently are not wetted by infiltrating water—the water merely moves through the spaces between fragments.

Exchangeable sodium percentage (ESP) is highly important in regulating infiltration rate and water storage of mine spoils in the Northern Great Plains. As ESP (the percentage of the cation exchange capacity occupied by Na) increases above about 12, spoils become dispersed and water entry is restricted. Runoff and erosion consequently increase, leaving less water stored in the soil for use by vegetation. The effects of sodic conditions are somewhat less pronounced in materials containing a high percentage of sand or appreciable amounts of biologically active organic matter. Restricted infiltration resulting from sodic conditions can be improved by replacing the exchangeable Na with Ca. This replacement can be done by adding gypsum (calcium sulfate or $CaSO_4 \cdot 2H_2O$) to the spoil material. Under the climate prevailing in most mine areas in the Northern Great Plains, gypsum additions reduce exchangeable Na content 30–50 percent in the upper 30 cm of material within a few years after treatment (Fig. 3.1). The Na replaced must be leached below the root zone, and this process can be accomplished by the use of mulches and fallow in

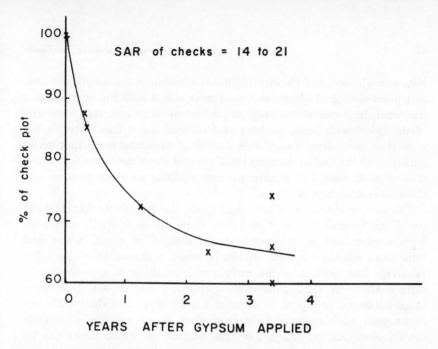

SAR of checks = 14 to 21

Figure 3.1   Reduction in Sodium Adsorption Ratio of North Dakota Mine Spoils Following Gypsum Treatment (for the range of values listed, SAR was found to be approximately numerically equivalent to ESP)

combination with gypsum. In more arid regions, natural precipitation may be too meager to accomplish this exchange and leaching of Na after gypsum treatment.

Sodium can be almost completely replaced within a few days by leaching with about 0.75 N calcium chloride ($CaCl_2$) (Doering and Willis 1975). This technique is known as high-salt leaching. But, because of the high concentration of soluble salts introduced, this treatment must be followed by leaching with 50–100 cm or more of irrigation water before plant growth is possible. Thus, this technique is expensive and requires good drainage and considerable supplemental water. Limestone or calcite ($CaCO_3$) is too low in solubility under alkaline conditions to be of use in replacing exchangeable Na.

Water infiltration and movement through spoils decreases as bulk density increases. However, water storage per unit depth increases with increased bulk density. Generally, bulk density of spoils is 10–30 percent lower than that of the original overburden. Typical bulk density values of overburden are 1.4–1.7 grams per cubic centimenter, whereas spoils are usually 1.1–1.4 gm/cm^3. Most spoils in the northern Great Plains are relatively dry at time of mining; therefore, use of heavy machinery seldom causes severe compaction. Sandy spoils may be an exception, and ripping or other surface disturbances of these spoils before seeding may be beneficial (Jensen and Hodder 1975). Overburden is commonly removed and piled with a dragline, and spoil piles are usually smoothed with a bulldozer. Both operations result in packing of spoils near the center axis of the piles, with lesser packing on the outslopes where materials are pushed or slide down slopes. Bulk density of smoothed spoils thus varies widely, with the higher densities usually found along the center line of the original spoil piles. This distribution may result in uneven subsidence and differences in surface drainage.

The steeper the slope of smoothed spoils, the greater the likelihood of runoff and erosion (Gilley et al. 1976). Many smoothed spoils have slope lengths over 100 m, permitting concentration of runoff water and subsequent erosion. On sodic spoils, however, sediment losses are often relatively low because of the surface seal, resulting in a pavementlike surface. However, loose sodic spoils are initially very erosive. Most states require that usable topsoil be saved and returned to smoothed spoils, but this topsoil readily erodes from long steep slopes, especially on areas devoid of vegetation (Table 3.6). The amount by which slopes can be reduced depends somewhat on the original topography of the mined area. If original slopes were steep and the topographic gradient of primary water courses was also steep, after mining smoothed spoils may have steep slopes or will need to be properly terraced. If the area was originally relatively level, postmining slopes can easily be reduced to grades of 5 percent or less.

Water Use

As important as improving infiltration and water storage in mined areas is enhancing the efficiency with which the stored water is used by productive vegetation. Under semiarid and arid conditions, essentially all of the available water within the root zone of a crop is used up by harvest time, regardless of the crop yield. Practices that increase plant growth and yield generally increase water-use efficiency under dryland conditions (Viets 1962). Soil fertility status, salinity level, thickness, compaction, and the presence of toxic materials all affect water-use efficiency.

Soil fertility is of major importance in regulating plant growth on mined lands. Proper control of soil fertility can greatly increase the efficiency with which water is used (Smika et al. 1965). Almost all mine spoils in the Northern Great Plains are deficient in plant-available P. Because most spoil materials in the West are calcareous, this deficiency is readily corrected, however, by applying 50–100 kilograms (kg) of fertilizer P per hectare (ha) before seeding the first crop. This quantity meets the requirements of grass for several years. Spoil materials are usually devoid of biologically active organic N materials, but many shales contain 10–50

Land Use	Runoff	Soil Loss
	mm	MT/ha
Rangeland	10	0.2
Spoil	51	7.8
Spoil covered with topsoil (25 cm)	41	36.0

1/ 9-10% slope, 100 mm water added in 4 hours to initially wet soil material.

Table 3.6 Runoff and Soil Loss from Rangeland Mine Spoils, and Spoils Covered with Topsoil, Using the Purdue Rainulator (Source: Gilley et al. 1976)

parts per million (ppm) exchangeable ammonium (NH_4) (Power et al. 1974). This substance is readily nitrified to NO_3-N within a few months after exposure to the atmosphere by the mining process. These shaley materials are frequently poorly drained, and the NO_3-N formed is often lost from the surface 25-50 cm, presumably by denitrification, within about two years (Table 3.7). Thus, available N in freshly exposed shale spoils is variable, but is usually low in older spoils. Until a pool of mineralizable organic N is built up, either annual fertilization or N-fixing legumes are required to sustain a supply of plant-available N in such spoils.

Potassium (K) has not been found to be deficient in mine spoils in the Northern Great Plains, just as it is seldom deficient for most dryland crops produced on unmined land. Predominant clay minerals are montmorillonites and illites, which hold considerable adsorbed K. Also, the sedimentary rock may contain some primary K-bearing minerals.

Seldom have deficiencies of other essential plant nutrients been recognized in mined spoils in the Northern Great Plains, although the amount of research on this subject to date is very limited. In some areas, exchangeable Mg levels are high enough to have the potential to restrict Ca uptake even from materials containing 10 percent free $CaCO_3$. Also, a few reports have indicated that molybdenum (Mo) levels in plants may occasionally be high enough to interfere with copper (Cu) nutrition of livestock. No selenium (Se) deficiencies or toxicities have been detected. Materials that contain appreciable biologically inert organic carbon (C),

Date[1]/	NH$_4$-N	NO$_3$-N
	ppm	ppm
May, 1971	17.1	21.4
August, 1971	2.1	28.7
August, 1975	1.6	10.5

[1]/ Spoils exposed by mining in March, 1971.

Table 3.7   Changes in Inorganic Nitrogen Content in Upper 30 cm of Unvegetated Mine Spoils Exposed to Normal Weathering

such as carbonaceous shales, leonardite, or coal slack, may sometimes contain toxic levels of boron (B), however.

High salinity reduces water-use efficiency by increasing the osmotic potential of the soil water. Consequently, the plant must expend more energy to extract water from the saline soil and the result is less growth per unit of water used. Only at a few locations in the Northern Great Plains is the level of total soluble salts high enough to interfere seriously with water uptake. Besides the osmotic effects of dissolved salts, individual salts may accumulate to toxic levels. For example, Mg may accumulate to the point that it interferes with Ca nutrition (Ries et al. 1976). The effects of Na accumulations were discussed earlier. At most locations, SO$_4$ is the predominate anion and Ca, Mg, and Na are the predominate cations in spoils (Sandoval et al. 1973). Salinity problems are commonly alleviated by leaching the soluble salts below the root zone, by means of either natural precipitation or supplemental irrigation. If leaching is not practical, then only salt-tolerant species can be grown on saline spoils.

As indicated, in dryland regions most plant species use essentially all the water available within the root zone by the end of the season. Therefore, efficient use of this water is enhanced not only by fertilization or other practices that stimulate growth, but also by proper selection of plant species. In the Northern Great Plains, perennial grasses are especially well adapted to the climate and soil resources that prevail. The proper selection of species for reclaimed land for optimum water utilization and for best support of domesticated livestock is critical.

Until the last decade or two, the prevailing philosophy was that only plant species native to the region should be considered for permanent sustained production. More recently, many other species have proved much more productive and well adapted to Northern Great Plains livestock ranching systems. It has also become apparent that both native and introduced grass species growing in favorable moisture sites respond well to fertilization. However, the use of introduced species, monocultures, fertilizers, and such practices requires intelligent management of livestock grazing to maintain the natural resources. In earlier generations, the ranching operations in the Northern Great Plains generally have not shown this degree of livestock management. (The effects of man's activities on grassland ecosystems in the West have been thoroughly discussed by Lewis [1969].)

RECLAMATION TECHNOLOGY

In the preceding paragraphs, many of the properties of mine spoils that interfere with soil-water relationships, and consequently restrict plant growth, have been identified, and potential methods of alleviating the adverse effects of these properties have been indicated. Unfortunately, many of these solutions are inadequate or impractical. In particular, changing the properties of sodic spoils is extremely difficult because present technology is extremely expensive, of limited effectiveness, or creates other problems.

Many of the problems encountered in reclamation can be solved by covering undesirable spoils with good soil material. A cover of as few as 5 cm of topsoil over sodic spoils increases the infiltration rate severalfold, reduces runoff, and vastly improves plant survival and growth (Sandoval et al. 1973). A thin layer of soil material absorbs raindrop impact, reducing surface sealing of sodic spoils. However, a cover of only 5 cm is of little value in enhancing water-holding capacity or the fertility status of the root zone. These are best improved by covering spoils with a greater thickness of acceptable soil material. In one experiment, first-year growth of spring wheat, alfalfa, crested wheatgrass, and native grasses all increased as total thickness of replaced soil (topsoil plus subsoil) spread over sodic spoils (SAR = 26) increased to about 70 cm (Table 3.8). Further increase in soil thickness had no consistent effect on production.

When high-quality topsoil was spread over the subsoil, yields with 70 cm or more of soil material were comparable to those obtained on unmined land. Applying topsoil and subsoil in separate layers is superior to mixing the two materials. With better quality spoils, containing fewer restrictions to growth, soil thickness requirements for maximum productivity may be less. On the other hand, thickness requirements may need to be increased somewhat to allow for salt migration upward from spoil into the replaced soil, for surface erosion losses, temporary waterlogging, or for other reasons.

Gilley and coworkers (1977) found that water erosion from sodic spoils

Topsoil	Subsoil Thickness, cm							
Thickness	10	30	50	70	90	110	130	150

A. Spring wheat grain yields, kg/ha

0	800	1062	1196	1290	1263	1250	1317	1250
20	1606	1915	1956	1942	1982	1949	2029	1922
60	1962	2016	2050	2050	1935	2009	2076	2130
Mixed[1]	1055	1344	1472	1505	1559	1478	1512	1452

B. Alfalfa (first cutting), kg/ha

0	108	446	805	970	829	691	834	869
20	720	637	1244	1152	1344	1287	1161	1207
60	912	930	980	1044	927	1323	1212	1085
Mixed[1]	128	380	808	952	1233	1057	1144	1217

C. Crested wheatgrass, kg/ha

0	1956	2498	2897	3039	3368	2769	3465	3194
20	2827	3194	3252	3697	3311	3137	3465	3165
60	2768	2942	3155	3251	3059	3280	2826	3097
Mixed[1]	1587	2498	3349	3252	3524	2980	3291	3368

D. "Native grass" mixture[2], kg/ha

0	17	151	118	181	320	272	339	296
20	463	609	871	1026	1048	864	790	656
60	325	133	491	432	366	250	260	314
Mixed[1]	0	7	84	165	103	85	345	349

[1] Topsoil and subsoil mixed 1:3 ratio; for other treatments topsoil spread over subsoil.

[2] Blue grama *(Bouteoua gracilis)* and side-oats grama (B. *curtipendula).*

Table 3.8   Yield of First Harvest of Several Crops Indicating Effects of Thickness of Subsoil and Topsoil Spread over Sodic (SAR = 26) Mine Spoils in North Dakota

covered with soil is much more severe than erosion from bare spoils. This condition arises from the fact that the pavementlike surface of sodic spoils is somewhat more resistant to the erosive action of water than is the loose replaced soil. For several reasons, most soil material is relatively dry when spread on spoils. Therefore, soil-covered spoils must be fallowed for several months to store sufficient water to insure germination and establishment of seedlings. During this time, these spoils are especially vulnerable to wind and water erosion. Research is in progress to determine the feasibility of adding a few centimeters of water after the soil material is spread, and seeding immediately afterward, with or without additional watering. Such an approach not only aids in solving the soil erosion problem, but also makes it possible to seed at almost any time during the growing season. The effects of supplemental watering on stand establishment, species competition, survival, and production are being investigated (Ries, Power, and Sandoval 1976). Information is also being acquired to determine water quality requirements and tolerances when only a few centimeters of water are applied to a seeding.

In recognition of the benefits of covering undesirable spoils with suitable soil material, legislation has been enacted in most major mining states requiring that suitable soil materials be saved and respread over graded spoils. In North Dakota, for example, before mining can begin, a soil survey of the permit area must be made, and overburden systematically sampled and analyzed. The topsoil (A horizon) is scraped off in accord with the soil survey, and is stockpiled. Next, the subsoil material (B and C horizons) is removed to a depth of 1.5 m, if present to that depth, and is also stockpiled. After mining and grading of spoils, the subsoil and then the topsoil materials are spread on the surface of the spoils. Thus practically all usable soil material is saved and returned, which, with proper fertilization and management, essentially ensures that potential productivity of the permit area will be equal to or greater than that before mining. If care is exercised in reduction of slopes of smoothed spoils, in some instances potential productivity after mining may be greater than that before mining. In areas where the supply of soil material is limited and spoils are sodic, application of gypsum before spreading the soil material may enhance water infiltration and root development into the soils.

CONCLUSIONS

Successful reclamation of surface-mined land in the Northern Great Plains is essentially the process of developing an environment conducive to the conservation and efficient utilization of precipitation by productive vegetation. Almost all potential types of postmining land use—agricultural production, grazing, wildlife, recreation, and so forth—are dependent upon successful revegetation. In the arid and semiarid regions concerned, successful revegetation is, therefore, essentially a problem of creating a plant-rooting medium that will effectively store precipitation where it falls, conserving the water in the plant root zone in a way that will enable productive plant species to use it most efficiently. If this objective is successfully accomplished, then the full productive potential determined by the available natural resources and local climate can be achieved on such land for this or future generations.

The conservation ethic dictates that the plant growth capability of the natural resources included in Northern Great Plains ecosystems not be diminished in the process of extracting the coal. It is not likely that mining activities will affect the resource of climate. Reclamation laws in most states now require that the soil resource be saved and returned to spoils after mining. As discussed earlier, preliminary research results have indicated that this could be done without appreciably diminishing the ability of the soil resource to produce vegetation. Disruption of both the groundwater and surface water resources will occur, but the latter can be restored and controlled by proper earth-moving activities during the mining process. The problem of restoring groundwater resources has not been solved. The vegetative resources can be duplicated and frequently improved in terms of potential use for man and animal, but we do not now have the technology to reproduce the native mixed prairie of the Northern Great Plains. Vegetation that can be established in its place will probably be more productive but will require more skillful management to maintain productivity.

The reclamation procedures currently required in most states appear adequate for restoration of plant growth potentials, but many problems that require additional research remain. Because these recently developed reclamation techniques have been used for only a few years, the long-term stability of the materials, of the landscapes and surface drainage, and of the perennial vegetation is largely unknown. Differential subsidence will probably continue for many years unless mining methods are changed to permit more uniform packing of spoil materials. Such subsidence will continually alter surface drainage. Problems relating to groundwater hydrology, which have not been addressed here, may be of major importance in some areas. Likewise, research is needed to restore wildlife habitats and other ecological niches. The need for or desirability of restoring native vegetation is another subject only briefly discussed here.

Much research remains to be done, but the outlook for restoring or enhancing the potential for producing vegetation useful to man appears promising, if the technology developed to date is properly utilized. Utilization of this technology can be achieved through proper enforcement of prudent legislation that recognizes the principles of the natural sciences involved.

REFERENCES

Averitt, P. 1975. *Coal Resources of the United States.* United States Geological Survey, Bulletin 1412.

Doering, E. J., and W. O. Willis. 1975. *Chemical Reclamation of Sodic Strip-mine Spoils.* USDA-ARS North Central Regional Publication No. 20.

Gilley, J. E., G. W. Gee, A. Bauer, W. O. Willis, and R. A. Young. 1977. *Runoff and Erosion Characteristics of Surface-mined Sites in Western North Dakota.* Transactions of the American Society of Agricultural Engineers, 20:697–704.

Jensen, I. B., and R. L. Hodder. 1975. *Effects of Surface Configuration in Water Pollution Control on Semiarid Surface Mined Land.* Billings, Montana: Proceedings of the Fort Union Coal Field Symposium, Montana Academy of Science.

Lewis, J. K. 1969. "Range Management Viewed in the Ecosystem Framework," in G. M van Dyne (ed). *The Ecosystem Concept in Natural Resource Management.* New York: Academic Press.

National Coal Association. 1975. *1974-1975 Coal Facts.* Washington, D. C.: National Coal Association.

Northern Great Plains Resources Program Staff. 1974. *Effects of Coal Development in the Northern Great Plains.* Denver: Northern Great Plains Resource Program.

Power, J. F., J. J. Bond, F. M. Sandoval, and W. O. Willis. 1974. "Nitrification in Paleocene Shale," *Science,* 183:1077-1079.

Ries, R. E., J. F. Power, and F. M. Sandoval. 1976. *Potential Use of Supplemental Irrigation for Establishment of Vegetation on Surface-Mined Lands.* N. Dak. Farm Res., 34:21–22.

Ries, R. E., F. M. Sandoval, J. F. Power, and W. O. Willis. 1976. "Perennial Forage Species Response to Sodium and Magnesium Sulfate in Mine Spoils," in *Proceedings of the Fourth Symposium on Surface Mining and Reclamation.* Washington, D. C.: National Coal Assocation.

Sandoval, F. M., J. J. Bond, J. F. Power, and W. O. Willis. 1973. *Lignite Mine Spoils in the Northern Great Plains—Characteristics and Potential for Reclamation.* Pittsburgh: Proceedings Research and Applied Technology Symposium on Mined-Land Reclamation.

Smika, D. E., H. J. Haas, and J. F. Power. 1965. "Effects of Moisture and Nitrogen Fertilizer on Growth and Water Use by Native Grass," *Agron. J.,* 57:483-486.

Viets, F. G., Jr. 1962. "Fertilizer and the Efficient Uses of Water," *Adv. in Agron.,* 5:223-264.

Part 2
Reclamation Studies at
Argonne National Laboratory

4

Assessment of Water Quality Impacts of a Western Coal Mine[1]

Edward H. Dettmann[2] and Richard D. Olsen[3]

Land Reclamation Program
Argonne National Laboratory
Argonne, Illinois

INTRODUCTION

Anticipated expansion of coal mining in many Western states implies a potential impact on water resources. The magnitude of the impact will be determined by the mining technology employed, and the associated hydrologic, meteorologic, and geologic characteristics of the mine locality.

Information describing historical and/or current aquatic impacts of Western coal mining is limited. It is clear, however, that because of fundamental regional differences in coal chemistry, as well as climatic and hydrologic differences, the environmental impacts to Western aquatic systems will differ significantly from those found in Eastern coal mining regions. At Western mines, acid drainage and the associated toxic metals, common in the East, should be minimal because of the small amounts of acid-forming substances (i.e., pyrite) and the generally alkaline nature of overburden and soils. Available research on Western mines indicates that leaching of soluble salts from mine spoils and transport of these salts to receiving surface or groundwater systems, as well as erosion-induced sedimentation, are among the principal water quality problems that can be expected (Van Voast 1974; McWhorter, Skogerboe, and Skogerboe 1975; Van Voast and Hedges 1975; McWhorter and Rowe 1976; Thurston, Skogerboe, and Russo 1976).

Also of concern in arid and semiarid regions are the potential adverse

53

impacts of mining on alluvial valley floors. The unconsolidated deposits often exist in a state of delicate hydrologic balance, which if upset could preclude future use of the area for agriculture and could also result in degradation of stream or groundwater quality (ICF 1976).

STUDY SITE, METHODS, AND MATERIALS

Study Area

This report describes interim results of a water quality investigation carried out during 1975–76 in the vicinity of the Big Horn Mine, an operating surface coal mine in the northwestern part of the Powder River Basin, Wyoming. The mine, which is located near Sheridan in the foothills of the Bighorn Mountains, has been in operation for approximately 20 years and is one of the several operating or proposed mines in the basin. Present coal production exceeds 1 million tons ($\sim$ 910,000 metric tons [MT]) per year.

The area is predominantly grassland with juniper and ponderosa pine present at elevations above about 4000 feet ($\sim$ 1200 meters [m]). The primary land use is grazing, but irrigated agriculture, principally hay production, is practiced in alluvial areas along perennial streams. Precipitation averages about 14 inches per year (36 centimeters [cm]/year); much of the total is snowfall. The mine site is traversed by two perennial streams, Goose Creek and the Tongue River, and the confluence is on mine property. Both streams have been diverted through the final cuts of past mining operations, thus forming two small lakes, one on each stream (see Fig. 4.1).

Two pits are being actively mined, the Zowada Pit east of Goose Creek and the Scott-Haymeadow Pit in the alluvial area south of the Tongue River. The rate of water seepage into the Scott-Haymeadow Pit adjacent to the Tongue River was approximately 3 cubic feet/second (cfs) (0.085 m³/s) during the study, primarily because of groundwater infiltration from the river through the alluvium into the pit. The seepage rate into the Zowada Pit was less than 1 cfs (0.028 m³/s) and was derived from flows through the coal seam at the highwall and seepage from surrounding spoil storage areas. Water from the Zowada Pit during 1975 and from the Scott-Haymeadow Pit during the entire term of the study was pumped to settling basins that drained through discharges 1 and 2 into Goose Creek and the Tongue River, respectively. During 1976, Zowada Pit discharge was pumped to a holding basin in the alluvium adjacent to Goose Creek (discharge 5). There was an additional discharge (3) directly from the Scott-Haymeadow Pit to the Tongue River during portions of this study.

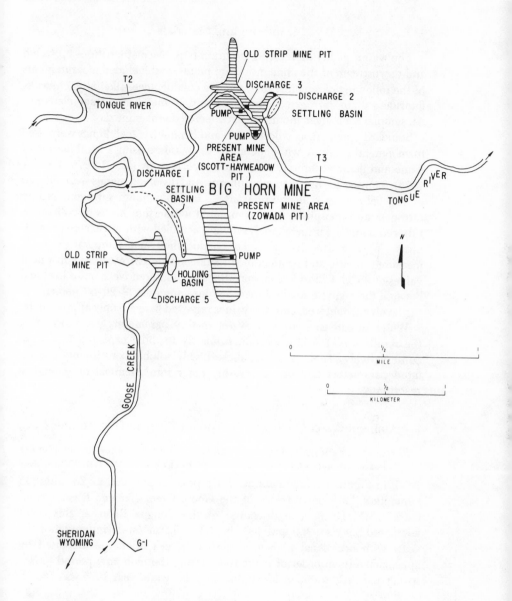

Figure 4.1　Map of Big Horn Mine and Vicinity

Methods and Materials

The water quality monitoring program took place at locations upstream and downstream of the mine discharge points, and included measurements of the following parameters: pH, specific conductance, alkalinity, chloride, fluoride, sulfate, nitrogen, phosphorus, and 16 metals and trace elements. All samples for a given month were collected on the same day.

Standard gravimetric, colorimetric and titrametric techniques were used for nonmetal analyses, while metals were measured using a combination of flame and flameless atomic absorption spectroscopy (USDI 1970; USEPA 1974). Samples for nitrogen, phosphorus, and dissolved metals were filtered in the field upon collection. Analyses for ammonium nitrogen, nitrate nitrogen, and phosphate were completed within four hours of collection. Filtered samples for metal analyses were acidified with nitric acid (5 milliliters [m] per liter [l]). Specific conductance and temperature were measured *in situ*, and conductance readings were corrected to equivalent values at 25°C. Chloride concentrations encountered were quite low, and because the analytic method used had low precision ($\sim$20–50 percent) at the levels encountered, chloride values reported here are only approximate.

While an extensive array of water quality parameters was monitored, the results reported here pertain primarily to those expected to behave conservatively (Table 4.1), i.e., those highly soluble constituents and related parameters that do not readily enter into chemical or biological reactions.

Ambient Water Quality in the Tongue River and Goose Creek

Seasonal variations in discharge and specific conductance in the Tongue River are summarized for water year 1975 (October 1974 to September 1975) in Figure 4.2. The data are for a point approximately 28 miles (45 kilometers [km]) downstream of the Goose Creek–Tongue River confluence (USDI 1975). The discharge of the Tongue River at this point fluctuated between 150 and 300 cfs (4.2 to 8.5 m³/s) during much of the year, with occasional higher flows. The river's discharge increased by approximately an order of magnitude during the high-flow period in late spring and early summer. Mean discharge for water year 1975 was 763 cfs (21.6 m³/s).

During this same period, specific conductance, a good index of total dissolved solids, fluctuated between 800 and 950 microohms/centimeter (μmhos/cm) during most of the year and decreased to approximately 250 μmhos/cm during the high-flow period, presumably because the dissolved solids are diluted by increased discharge during snowmelt.

The same seasonal patterns of discharge and specific conductance hold for the Tongue River and Goose Creek upstream of their confluence. The mean discharge of Goose Creek at the U.S. Geological Survey (USGS)

Parameter	Units	Stations			Ratio
		G1	T2	T3	G1/T2*
Specific Conductance @ 25°C	$\left(\dfrac{\mu\text{mhos}}{\text{cm}}\right)$	615	349	451	1.8
Total Dissolved Solids	(mg/ℓ)	552	320	412	1.7
Bicarbonate	(mg/ℓ)	312	207	246	1.5
Chloride	(mg/ℓ)	2.92	0.71	1.99	4.1
Fluoride	(mg/ℓ)	0.36	0.19	0.27	1.9
Sulfate	(mg/ℓ)	147	58	95	2.5
Calcium	(mg/ℓ)	49	40	40	1.2
Magnesium	(mg/ℓ)	40.9	19.1	28.4	2.1
Sodium	(mg/ℓ)	27	12.6	17.8	2.1
Potassium	(mg/ℓ)	2.7	1.3	1.8	2.1

*These ratios are dimensionless.

Table 4.1 Mean Concentrations of Conservative Parameters in Goose Creek and the Tongue River

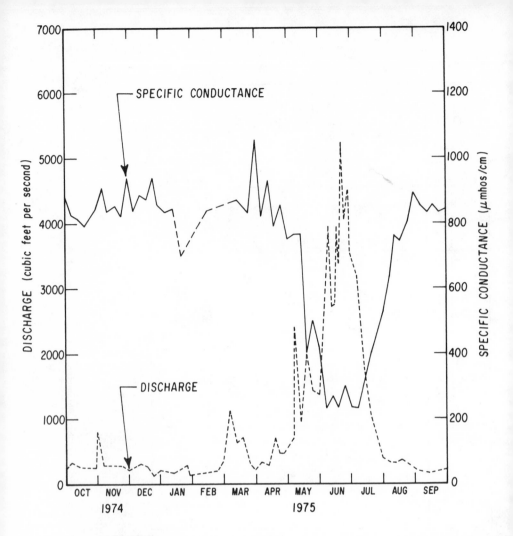

Figure 4.2 Seasonal Discharge and Specific Conductance of the Tongue River and the Wyoming-Montana State Line During Water Year 1975 (Source: USDI 1975)

gaging station below Sheridan, approximately 11 miles (18 km) upstream of the confluence, was 259 cfs (7.33 m³/s) in water year 1975. Monthly instantaneous discharge measurements for a USGS gaging station at Monarch on the Tongue River, approximately 4 miles (6.4 km) upstream of the confluence, averaged 454 cfs (12.9 m³/s) for water year 1975. For Goose Creek on the same dates, the mean discharge was 359 cfs (10.2 m³/sec).

Mean concentrations of conservative water quality parameters in the Tongue River and Goose Creek are summarized in Table 4.1. Stations G1 and T2 are in Goose Creek and the Tongue River, respectively, upstream of their confluence; station T3 is downstream of the confluence (see Fig. 4.1). The concentrations of most ions are between 1.2 and 2.5 times higher in Goose Creek than in the Tongue River at station T2. As would be expected, concentrations for all conservative parameters at the downstream station T3, representing the combined flow of both streams, lie between the values for the separate streams. These data indicate that Goose Creek has a large effect on the water quality of the Tongue River.

Water Quality of Pumped Mine Discharges

Water quality data for the three pumped mine discharges and Tongue River station T3 are shown in Table 4.2. Sample standard deviations are included with mean concentrations to indicate variability.

Table 4.2 includes data on pH as well as conservative parameters. The pH values for the discharges are near 8, approximately equal to those for the Tongue River. This alkaline drainage is characteristic of Western coal mines, whereas acid effluents are often found at mines for high-sulfur Midwestern and Eastern coals.

Concentrations of all conservative parameters are highest in D1 and D5. All constituents are present at lower levels in D2 and D3.

Ratios of the concentrations in the pumped mine discharges to the concentrations in the Tongue River are summarized in Table 4.3. Ions are most concentrated in D1 and D5 and least in D2. In all discharges, concentrations of all constituents listed exceed those in the Tongue River.

While concentrations of most parameters are elevated above ambient levels by a factor of 1.1 to 6.8 for D1 and D5, sulfate and potassium are elevated by a factor of 15 and sodium by a factor of 23. These three ions are also higher in ambient concentrations than other ions in the other two discharges. Concentration ratios for D2 and D3 are, however, considerably smaller than those for D1 and D5. A large fraction of the effluents from D2 and D3 consists of water seeping from the Tongue River through alluvial and/or spoil material into the Scott-Haymeadow Pit adjacent to the river. This river water is dilute compared with mineralized water entering the pits from other sources.

Effect of Mining on Stream Water Quality

The concentrations of dissolved substances downstream of the mine and confluence are determined by: their upstream concentrations in Goose Creek and the Tongue River, their concentrations in the inputs to the stream system, and the volumetric flow rates of these sources. External inputs to the streams may consist of surface runoff and groundwater

| Parameter | Units | Discharges | | | Tongue River |
		D1 and D5	D2	D3	T3
pH		8.06 (0.23)	7.98 (0.13)	7.78 (0.09)	8.35 (0.23)
Specific Conductance @ 25°C	($\frac{umhos}{cm}$)	3078 (384)	673 (104)	1098 (158)	451 (161)
Total Dissolved Solids	(mg/ℓ)	2730 (394)	606 (41)	915 (105)	412 (158)
Bicarbonate	(mg/ℓ)	554.3 (46.7)	304.5 (13.5)	329.1 (39.6)	246.0 (66.0)
Chloride	(mg/ℓ)	12.0 (5.5)	2.24 (0.79)	5.14 (1.27)	1.99 (1.48)
Fluoride	(mg/ℓ)	0.79 (0.35)	0.43 (0.17)	0.53 (0.32)	0.27 (0.15)
Sulfate	(mg/ℓ)	1411 (266)	166 (23)	363 (81)	95 (44)
Calcium	(mg/ℓ)	161 (45)	53 (6)	83 (18)	40 (15)
Magnesium	(mg/ℓ)	138 (22)	41 (4)	65 (12)	28 (12)
Sodium	(mg/ℓ)	406 (78)	34 (4)	55 (5)	18 (9)
Potassium	(mg/ℓ)	26.2 (4.9)	5.2 (0.4)	7.8 (1.1)	1.8 (0.7)
Sample Size		6 to 8	5 to 9	4 (3 for fluoride)	9 to 10

Table 4.2   Mean Concentrations and Sample Standard Deviations (in Parentheses) for Conservative Parameters and pH in Pumped Mine Discharges and the Tongue River

Parameter	Concentration Ratios		
	D1/T3, D5/T3	D2/T3	D3/T3
Specific Conductance @ 25°C	6.8	1.5	2.4
Total Dissolved Solids	6.6	1.5	2.2
Bicarbonate	2.3	1.2	1.3
Chloride	6.0	1.1	2.6
Fluoride	2.9	1.6	2.0
Sulfate	15.	1.8	3.8
Calcium	4.0	1.3	2.1
Magnesium	4.9	1.5	2.4
Sodium	23.	1.9	3.1
Potassium	15.	2.9	4.3

Table 4.3 Ratios of Concentrations of Conservative Parameters in Pumped Mine Discharges to Concentrations in the Tongue River at Sampling Station T3

seepage unrelated to mining, as well as mine-related sources, e.g., the pumped discharges described above, surface runoff, seepage from spoils, and seepage from aquifers such as coal seams, which come into hydraulic contact with the streams as a result of mining. Examples of the last category are the aquifers exposed in the mine pits through which Goose Creek and the Tongue River have been diverted.

The total loading rate of any given parameter to the stream reach downstream of the confluence is given by the expression $C_t D_t + C_g D_g + \sum_i C_i D_i$, where:

C_t = observed concentration of the parameter upstream on the Tongue River,

C_g = observed concentration of the parameter upstream on Goose Creek,

C_i = observed concentration of the parameter in the i'th external input,

D_t = upstream flow rate of the Tongue River,

D_g = upstream flow rate of Goose Creek, and

D_i = flow rate of the i'th external input.

If one assumes complete mixing and conservative behavior for the parameter of interest, the mass balance can be expressed as:

$$C_d = \frac{C_t D_t + C_g D_g + \sum_i C_i D_i}{D_t + D_g + \sum_i D_i} \qquad (1)$$

where C_d is the calculated concentration sufficiently far downstream of the confluence and the inputs to guarantee complete mixing. The sum of discharges in the denominator gives the downstream discharge of the Tongue River.

Although some quality and flow data are available for pumped mine discharges, the loading rates from other sources such as spoil pile seepage and groundwater seepage directly into the streams through the pit walls are presently unknown. For this reason, the approach used in assessing the effect of mining was to calculate anticipated downstream concentrations solely on the basis of ambient upstream water quality in Goose Creek and the Tongue River using equation 2 below, and to compare these expected concentrations with those observed at station T3.

$$C_d = \frac{C_t D_t + C_g D_g}{D_t + D_g} \qquad (2)$$

Calculated and observed concentrations would be expected to agree closely in the absence of any significant water quality impacts of the mine. Observed concentrations of conservative substances consistently exceeding calculated values would indicate that a source of these materials exists in the vicinity of the mine. This procedure is an extension of the commonly used technique of comparing concentrations of water quality parameters upstream and downstream of a point source. The method permits assessment of effects of multiple sources in the complex stream system encountered at the site.

The calculated and observed values of specific conductance and concentrations of total dissolved solids, as well as eight individual ions expected to behave conservatively in this system, are compared for eight sampling dates between August 1975 and November 1976 (Table 4.4). Data for two sampling dates in June of 1975 and 1976 were not included in the analysis because substantially reduced concentrations of all ions during the high-flow period are atypical. Also, the extremely high stream flows during this period are expected to dilute mine effluents more than usual. The effect would be the biasing of conclusions by inclusion of data collected when the system is particularly insensitive to the effects of

Parameter	Mean Deviation $(C_{obs} - C_{calc})^{*,**}$	Standard Error of Mean**	Mean Deviation / Standard Error	Sample Size
Specific Conductance @ 25°C	-2.31	9.98	-0.23	8
Total Dissolved Solids	+0.11	9.11	0.012	7
Bicarbonate	-4.86	2.79	-1.7	8
Chloride	+0.577	0.225	2.6	5
Fluoride	+0.005	0.018	0.28	7
Sulfate	+2.091	5.783	0.36	8
Calcium	-3.619	2.715	-1.3	8
Magnesium	+0.758	0.946	0.80	8
Sodium	-0.363	0.623	-0.58	7
Potassium	-0.071	0.109	-0.65	7

*Observed concentration at Station T3 minus calculated concentration.

**Units for specific conductance are μmhos/cm, units for all other parameter are mg/ℓ.

Table 4.4   Mean Deviations Between Observed and Calculated Concentrations at Tongue River Sampling Station T3, and Standard Errors of the Means

mining. An analysis that included these additional data points did not, however, alter the conclusions described below.

The data in the first column of Table 4.4 give the mean of the differences between observed concentrations at sampling station T3 and the concentrations calculated using equation 2. Deviations from zero difference could be caused by stream loading (e.g., mine effluents) between the upstream and downstream sampling sites and/or by measurement variations within the levels of analytical precision. An additional source of possible deviations is time-dependent changes in ambient concentrations accompanied by time delays in transport, particularly in the two pits through which the streams are diverted. Because deviations of this sort are expected to be random, mean deviations attributable to them should be small and could be either positive or negative. Loading by mine effluents or other sources between the upstream and downstream sampling points would lead to observed concentrations at station T3 in excess of those calculated, giving a positive mean deviation. Sulfate, sodium, and potassium concentrations may be particularly sensitive indicators of mining impacts because these ions are most elevated in mine effluents.

The second and third columns of Table 4.4 contain the standard errors of the mean deviations and the mean deviations divided by the standard errors of these mean deviations. Use of a t test (one- or two-tailed) indicates that all of the mean deviations except that for chloride are consistent at the 5 percent significance level with the null hypothesis that the mean deviations are drawn from a distribution with mean zero. In particular, the mean deviations of sulfate, sodium, and potassium, the elements most concentrated in pumped mine discharges, are within 0.36, 0.58, and 0.65 standard deviation of zero.

We conclude that any effect of mining on conservative ions other than chloride during the study was within the range of sampling and analytical precision (estimated at ~10 percent) and of short-term variations in ambient water quality. Thus, any possible effect is undetectable using the available data. In view of the low precision of the chloride determination and the fact that chloride is not highly concentrated in mine effluents when compared with sulfate, sodium, and potassium, the same conclusion is probably applicable to this ion as well.

Upstream Water Quality Survey

To place potential water quality changes in the vicinity of the Big Horn Mine in perspective, in the late summer of 1975, a survey of spatial variations in specific conductance was conducted in the Tongue River and Goose Creek watersheds, beginning at the base of the Bighorn Mountains. The specific conductance of Tongue River water rose from a value of 219 μmhos/cm at the base of the Bighorn Mountains to a value of 377 μmhos/cm just upstream of the mine. This change occurred over a distance of approximately 25 stream miles (40 km).

Specific conductance changed more dramatically in Goose Creek and its tributaries, Little Goose Creek and Big Goose Creek, which join to form Goose Creek in the city of Sheridan. The specific conductance in Little Goose Creek, for example, rose by a factor of 5.7, from a value of 91 μmhos/cm near the foot of the Bighorn Mountains to 519 μmhos/cm in a distance of only 5 miles (8 km). Similar large changes were observed in Big Goose Creek. Specific conductance continued to rise in Goose Creek to values near 900 μmhos/cm upstream of the Big Horn Mine. U.S. Geological Survey data suggest that bicarbonate, sulfate, calcium, magnesium, and sodium ions are largely responsible for these increases (USDI 1975). Data collection during the next field season will be devoted in part to isolating the causes of these increases in specific conductance. One possible source is irrigation return flow, since irrigation is widely practiced in the watershed.

Summary and Discussion

The effect of the Big Horn Mine on concentrations of dissolved conservative constituents in the Tongue River is small, and is within the range of analytical precision and short-term variations in ambient concentrations. On the other hand, large changes in stream water quality are evident in upstream reaches of the Goose Creek and Tongue River watersheds, where intensive agricultural activity exists.

Trace element concentrations are relatively low at all stream and mine discharge sampling points (Olsen and Dettmann 1976), but these data are preliminary, and not definitive at this stage of the study. Ammonium and nitrate concentrations are elevated in the Big Horn Mine discharges (possibly because of use of ammonium nitrate explosives), but phosphorous concentrations are met. Material balance calculations indicate that present loading by pumped mine discharges from the Big Horn Mine could increase ammonium and nitrate concentrations locally in the Tongue River by approximately 3 percent and 1 percent, respectively. While this increase would probably represent no measurable adverse impact, the cumulative effect of nitrogen loading at expanded mining levels should be investigated—particularly the potential for eutrophication in the Tongue River.

Van Voast and Hedges (1975) have estimated the potential water quality impacts of expanded coal extraction at the Decker Mine on the Tongue River. This mine is located adjacent to the Tongue River Reservoir, approximately 30 stream miles (48 km) downstream of the Big Horn Mine, and has a yearly production of approximately 10 million tons (~ 9.1 million MT). It is currently the largest operating surface coal mine in the U.S. Worst-case estimates indicate increases in the range of 0.–3.5 percent above ambient levels for the sodium adsorption ratio and for concentrations of total dissolved solids, sodium, calcium, and magnesium in the Tongue River during mining, and increases in the range 0.–7.4 percent during the postmining period.

Studies conducted at the Edna Coal Mine near Oak Creek, Colorado, found substantial (often severalfold) increases in specific conductance and total dissolved solids for water in a reach of Trout Creek receiving mine effluents (McWhorter, Skogerboe, and Skogerboe 1975; McWhorter and Rowe 1976). The authors present evidence that a large fraction of the increase is attributable to drainage from areas disturbed by mining. The primary ions in runoff from the Edna Mine site are calcium, magnesium, bicarbonate, and sulfate. The hydrologic regime of this site differs significantly from that of the Tongue River sites. Mean annual precipitation exceeds 20 inches (51 cm), and runoff from the mined area represents a much larger fraction of the total stream discharge than is the case at the Big Horn or Decker sites.

The results obtained by McWhorter et al. at the Edna Mine indicate that salinity increases may cause substantial water quality impacts at some Western coal mines. The magnitude of such impacts is, however, sensitive to site-specific hydrologic conditions. Our study and that of Van Voast and Hedges at the Decker site suggest that the current mining (i.e., 1975–76) at the Big Horn and Decker mines and the proposed Decker expansion will have only a minor effect on salinity in the Tongue River. The small effects observed at the Big Horn and Decker sites appear in large part attributable to the large quantities of dilution water available in the Tongue River relative to mine discharges.

In view of anticipated expansion of energy development in the Tongue River watershed, it should be stated that this conclusion applies only to mining at the indicated levels. Future water quality effects of expanded coal mining and utilization on the Tongue River will depend on both the quantity and quality of energy-related effluents released into the river, and upon perturbations such as consumptive water use and aquifer disturbance. Operation of additional mines along the Tongue River could possibly cause measurable increases in salinity and other parameters.

NOTES

1. Work performed under the auspices of the U.S. Energy Research and Development Administration. The study was undertaken in coordination with Sheridan Community College and in cooperation with Peter Kiewit and Sons, Coal Mining Division.

2. Division of Environmental Impact Studies, Argonne National Laboratory.

3. Acknowledgment: the authors wish to thank Dr. Roger Ferguson and Mr. Jay Lance of Sheridan Community College for aid in collection of samples and for nonmetal laboratory analyses.

REFERENCES

ICF Incorporated. 1976. *Impacts of Alluvial Valley Floor Provisions in H.R. 13950.* Draft submitted to Council on Environmental Quality, November 1976.

McWhorter, David B. and Jerry W. Rowe. 1976. *Inorganic Water Quality in a Surface Mined Watershed.* Paper presented at: American Geophysical Union Symposium on Methodologies for Environmental Assessments in Energy Development Regions, December 8, 1976, San Francisco, California.

McWhorter, David B., Rodney K. Skogerboe, and Gaylord V. Skogerboe. 1975. *Water Quality Control in Mine Spoils, upper Colorado River Basin.* Cincinnati, Ohio: Report EPA–G70/2–75–048, U.S. Environmental Protection Agency.

Olsen, Richard D. and Edward H. Dettmann. 1976. "Preliminary Results from a Study of Coal Mining Effects on Water Quality of the Tongue River, Wyoming," *Hydrology and Water Resources in Arizona and the Southwest,* 6:59–67.

Thurston, Robert V., Rodney K. Skogerboe, and Rosemarie C. Russo. 1976. *Toxic Effects on the Aquatic Biota from Coal and Oil Shale Development: Progress Report—One Year.* Fort Collins, Colorado: Internal Project Report No. 9, Natural Resource Ecology Laboratory, Colorado State University.

USDI. 1970. "Techniques of Water-Resource Investigations of the U.S. Geological Survey," *U.S. Geological Survey Book* A5.

USDI. 1975. *Water Year 1975, Water Resources Data for Wyoming.* Cheyenne, Wyoming: U.S. Geological Survey.

USEPA. 1974. *Methods for Chemical Analysis of Water and Wastes.* Cincinnati, Ohio: EPA–625/6–74–003, U.S. Environmental Protection Agency.

Van Voast, Wayne A. 1974. *Hydrologic Effects of Strip Coal Mining in Southeastern Montana—Emphasis: One Year of Mining Near Decker.* Montana Bureau of Mines and Geology, Bulletin 93, June.

Van Voast, Wayne A. and Robert B. Hedges. 1975. *Hydrologic Aspects of Existing and Proposed Strip Coal Mines near Decker, Southeastern Montana.* Montana Bureau of Mines and Geology, Bulletin 97.

5

Economics of Mined Land Reclamation and Land-Use Planning in Western States[1]

James R. LaFevers

Energy and Environmental
Systems Division
Argonne National Laboratory
Argonne, Illinois

The cost-effectiveness of reclamation programs in mined areas is closely related to an integrated approach to extraction, reclamation, and land-use planning. Prior to mining, a plan should be developed that includes extractive processes and reclamation techniques designed to create a landscape that will be of benefit to the local community or region in which the mine is located. If the reclaimed land can be used to satisfy a local or regional land-use need, it will be of much greater value than land that is reclaimed to approximate premining conditions without consideration of marketability or future community needs. An important aspect of land-use planning at mine sites is that it assumes that the condition of the land prior to mining is not necessarily the optimum condition to which it can be reclaimed. In actuality, many surface-mining operations present opportunities for improvement over premining conditions.

For the mining industry to create landscapes that will be of the greatest benefit to the community, inputs from local or regional land-use planners are needed. In many cases, public sector planners could provide information on future land-use needs, development criteria, and other factors that could be used in developing marketing strategies for reclaimed sites.

Traditionally, however, such inputs have not been forthcoming. Neither the mining industry nor the planning community has been effective at initiating interaction with the other. For the most part, industry views reclamation as an unprofitable undertaking and has avoided the public planning community for fear that interaction might become mandatory. The planners, on the other hand, have generally ignored mining operations until after the fact. Numerous abandoned mine sites have been incorporated into revised land-use plans, for example. It is much more difficult, however, to find sites at which public sector planners made inputs to plans for postmining use of the site before the mining was conducted. This lack of input is partly because planners have no mandate to become involved in such activities, partly because they have a very poor relationship with industry, and partly because they do not traditionally have the background for creating new landscapes *ex nihilo.*

Many state laws and the federal reclamation bills have provisions for reclaiming mined areas to premining contour and natural vegetation. In many cases, such requirements could preclude effective land-use planning. The potential benefits from integrated reclamation and land-use planning vary according to several factors. Physical factors—such as climate, soils, and the nature of the disturbance caused by extraction—vary with location and the commodity being mined. In addition, the costs of reclamation and the requirements of reclamation laws can limit the planning potential. The potential for reuse of a mined site is also related to certain factors that vary regionally. Population pressure is one of the most significant factors. In a general model, the higher the population pressure, the greater the land-use needs. The greater the land-use needs, the greater the demand, or competition for the land. As competition increases, the land values also increase, and so does the potential for land-use planning. In high density population areas, where land values are generally high and land-use needs are varied, a potential might exist for the creation of recreational, residential, industrial, commercial, or other intensive use sites. In such cases, the value of the reclaimed site may exceed the cost of reclamation by a considerable margin. The reclamation, in addition to being beneficial to the mining company, also benefits the public, whose land-use need is partially alleviated. Because of the population factor, the relatively sparsely populated arid states have a lower reclamation/land-use planning potential than the more populous areas.

We examined a number of mine sites in the arid and semiarid states of the Western U.S. At a copper mining operation in Nevada, where no reclamation was planned or conducted, reclamation costs were zero; the benefits from reclamation were also zero. The state has no reclamation law, and because it has not generally been demonstrated that reclamation of the type of site encountered can be profitable, no future use of the site is planned. An abandoned gold placer mine near the copper mine indicated the type of self-regeneration of vegetation that could be expected. Spoils that were more than 100 years old still showed no vegetation whatever.

Another state that has no reclamation law is Arizona. A number of copper mines near Tucson were examined. The rock waste and tailings disposal problems were the same at these sites as at the Nevada site. Revegetation projects have, however, been initiated at most of the Arizona sites examined, probably because these sites are very close to the city of Tucson and to the retirement community of Green Valley. For public relations purposes, and possibly because of the potential future demand for land and related increases in land values, a number of reclamation research programs and revegetation projects are under way at these sites. These projects consist primarily of the establishment of vegetation on the outer slopes of tailing pond berms, although at some sites reclamation plans call for the eventual covering and revegetation of the surfaces of tailings and waste rock piles. At present, most revegetation includes hand planting of native bushes and shrubs, and exotic grasses and trees. Irrigation is used to establish growth. Reclamation costs are estimated at approximately $2000 per acre, with about one-half the cost being for the irrigation system. Average land values in the region are from $100 to $500 per acre, but could be much higher for specialized uses.

In northern Arizona, on the Black Mesa, a coal mining operation was examined. Premining vegetation is extremely sparse and commercially poor, primarily as a result of extreme overgrazing. The region, the Navajo-Hopi joint use area, is used for sheep and cattle grazing. Current land values are indeterminate. After mining, the land is returned to range conditions, with the animal carrying capacity improved over premining conditions. Reclamation costs, which must include fencing to keep the herds out until vegetation is properly established, are about $5000 per acre.

Four lignite mining sites were examined in North Dakota. Reclamation costs averaged about $2400 per acre through 1975—including grading, backfilling, seeding, soil amendments, and the removal and replacement of up to 18 inches of plant growth material (topsoil). Backfilling and grading costs were 55 percent of the total reclamation cost. Topsoiling costs averaged about 28 percent of the total. Recent changes in the state reclamation law have made necessary the removal, segregation, and replacement of up to 5 feet of topsoil. As a result of the new requirements, reclamation costs are now exceeding $6600 per acre at some sites. Backfilling and grading are no longer the most expensive reclamation procedures. The topsoiling process now accounts for approximately 61 percent of the total reclamation cost. Backfilling and grading have dropped to 38 percent. The value of unmined adjacent land is approximately $200 per acre. Although reclamation costs can run as much as 30 times the value of the land, in terms of cost per million British thermal units (Btu), reclamation costs vary from 6 to 20 percent of the value of the lignite that is mined.

NOTE

1. This paper is *condensed* from a presentation at the 143rd Annual Conference of the American Association for the Advancement of Science.

6

An Analysis of the Sensitivities of Local Socioeconomic Impacts to Variations in the Types and Rates of Coal Development and Differences in Local Site Characteristics[1]

Erik J. Stenehjem
L. John Hoover
Gregory C. Krohm

Energy and Environmental Systems Division
Argonne National Laboratory
Argonne, Illinois

INTRODUCTION

Numerous studies have been conducted to predict the socioeconomic effects associated with site-specific coal development projects. The generic nature of most socioeconomic studies has been to forecast the employment and population impacts accompanying a specific energy technology at a particular site.such forecasts have, in fact, become an important component of the environmental impact analyses of site-specific coal extraction and conversion activities. When these studies are carefully prepared, they provide government, industry, and interest groups with important information about the expected nature of social and economic changes at a particular site. Such studies of the socioeconomic impacts of a particular coal technology at a particular site are, however, of little value in the evaluation and comparison of alternative sites, or in determining regional patterns of development. In order to assist energy planning officials in both the public and private sectors in the evaluation of sites, patterns of development, and coal technologies that will minimize social and economic impacts, a more general analysis is required. This paper represents an initial attempt to provide such an analysis. It is intended to serve as a basis for discussion so that subsequent analyses can more thoroughly address the issues. As a part of the Regional Studies Program being sponsored by the Administrator for Environment and Safety of the Energy Research and Development Administration (ERDA), Argonne and

72

several other national laboratories are conducting a National Coal Utilization Assessment (NCUA) to identify the problems associated with increased coal use and to analyze alternative mitigation strategies. This paper will serve as background for the Midwest regional assessment to be completed in July 1977, and the national assessment in December 1977.

Using an approach developed at Argonne for use in its Regional Studies Program, the socioeconomic consequences accompanying alternative rates and types of coal development in a variety of potentially exploitable counties are projected. The results of these "mini" case studies are then used to evaluate the sensitivity of the projected impacts to the factors outlined above. By contrasting the socioeconomic consequences of identical coal technologies in counties having markedly different economic, geographic, and demographic characteristics, the influences of preexisting conditions and infrastructure on the rate and magnitude of such impacts are described. Alternatively, the influences of different rates and types of coal development are evaluated by contrasting their anticipated socioeconomic consequences among nearly identical counties.

COAL PRODUCTION AND USE PATTERNS

Under any reasonable set of conditions, domestic coal utilization is likely to increase considerably above recent demand patterns. Within most regions of the United States, coal demand has been either declining or growing only slightly for the last two decades. From 1955 to 1975, domestic coal demand grew at an annual rate of about 1.5 percent, compared to an overall energy demand growth of nearly 3.5 percent. As a result of this lasting demand, coal currently supplies less than one-fifth of U.S. energy needs.

Coal's historical pattern of decline has been arrested by the sudden surge in demand following the oil embargo of 1974. U.S. production of bituminous coal and lignite reached the record production level of 665 million tons in 1976. U.S. coal production and demand are forecast to increase to over 1 billion tons per year by 1985. This amount represents a doubling of coal output in less than 10 years. An even greater growth in demand will occur if synthetic fuels achieve widespread commercialization. These future demand patterns, shown by region in Table 6.2, can be contrasted with the present pattern in Table 6.1. Throughout the period under consideration, electric utilities remain the dominant source of demand (particularly in the midcontinent regions). Industrial demand is the second largest category, but by the turn of the century, synfuel feedstock is approximately one-half of total industrial consumption of coal.

In addition to a fairly rapid change in overall demand for coal, the

Region	Utility	Industrial	Synfuel	Res/Com
New England	2.6	.3	0	.01
Middle Atlantic	55.4	23.5	0	.30
East North Central	159.2	72.0	0	2.49
West North Central	34.3	9.0	0	.47
South Atlantic	81.2	28.1	0	.83
East South Central	66.7	14.0	0	.42
West South Central	3.7	.7	0	*
Mountain	20.5	3.6	0	.27
Pacific	1.2	.7	0	.02
	424.9	151.9	0	4.81

*Denotes negligible amount

Table 6.1   Domestic Coal Demand by Census Region (Millions of Tons/Year) 1975

nation is likely to witness an even more dramatic shift in coal production from traditional mining areas in the East and Midwest to relatively unexploited coal fields in the West. Figure 6.1 depicts these coal regions, and Table 6.3 illustrates the magnitude of this shift. The major impetus to this shift is not the increase in the general demand for coal, but the restrictions placed on the combustion of high-sulfur coal.

Although coal mining districts in Appalachia and the Interior coal provinces have extremely large resource deposits of a generally superior quality, most of this coal has a high-sulfur content. While sulfur levels vary significantly from county to county, many of the largest Eastern and Interior deposits contain sulfur levels in the range of 3.0–3.5 percent. This volume of sulfur is difficult to control with current flue gas desulfurization technologies. Moreover, the current cost of these controls is prohibitive.

The expense and uncertainty of sulfur control technologies have created a new and potentially expansive market for low-sulfur Western coals in the high demand areas east of the Mississippi. Figure 6.2 illustrates the

	1985			
			Synfuel	
Region	Utility	Industrial	Feedstock	
New England	6.5	0.3	.07	*
Middle Atlantic	110.6	47.4	3.60	0.16
East North Central	109.8	100.5	6.75	1.38
West North Central	66.1	9.6	1.50	0.26
South Atlantic	143.0	33.4	0.50	0.46
East South Central	79.6	24.1	0.90	0.23
West South Central	45.4	2.0	0.80	*
Mountain	43.3	6.6	1.20	0.15
Pacific	10.7	3.2	4.75	0.01
TOTAL	695.3	227.1	20.07	2.67

	2000			
New England	5.6	*	3	*
Middle Atlantic	159.9	64.0	32	0.07
East North Central	259.9	145.0	47	0.57
West North Central	104.3	16.0	21	0.11
South Atlantic	159.1	55.0	20	0.19
East South Central	84.4	40.0	20	0.10
West South Central	60.1	3.0	14	*
Mountain	52.3	9.0	10	0.06
Pacific	16.9	5.0	14	*
TOTAL	902.5	337.0	181	1.10

*Denotes negligible amount

Table 6.2   Domestic Coal Demand by Sector (Millions of Tons/Year)

probable fringe of Eastern market penetration by Western low-sulfur coal. States west of the shaded area have already made substantial commitments to supply their coal-burning electrical generating capacity with Western coal. Industrial demand for low-sulfur coal is also expected to rise sharply in this general market area.

The Northern Great Plains will be the dominant source of this Western coal supply. The states of Montana, Wyoming, and North Dakota have enormous deposits of strippable coal. The North Dakota lignite, though plentiful and inexpensive to mine, is of extremely low quality. For this

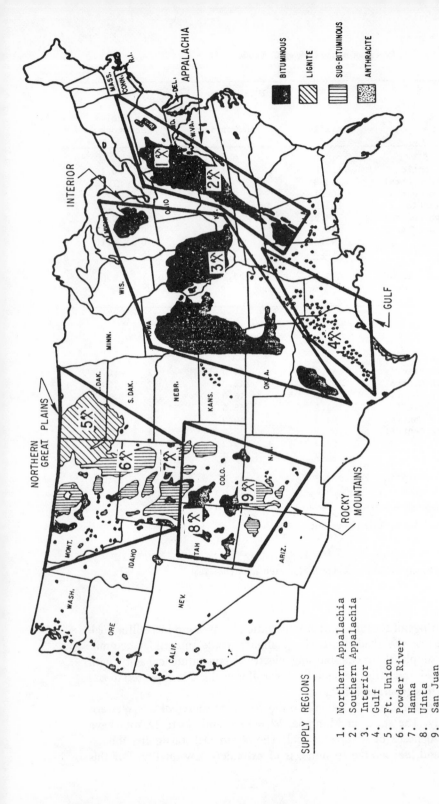

SUPPLY REGIONS

1. Northern Appalachia
2. Southern Appalachia
3. Interior
4. Gulf
5. Ft. Union
6. Powder River
7. Hanna
8. Uinta
9. San Juan

BITUMINOUS
LIGNITE
SUB-BITUMINOUS
ANTHRACITE

Figure 6.1 Coal Provinces

Appalachia	392	568	672
Interior	147	160	216
Gulf	11	25	57
Northern Great Plains	55	199	471
Rocky Mountains	35	73	143
U.S. Total	640	1,025	1,559

Table 6.3   Coal Production by Major Province (Millions of Tons/Year)

reason it will probably not be exported in its natural state but instead will be converted to either electrical power or synthetic natural gas (SNG). The major source of western coal "exports" to Eastern markets will center on the Powder River Basin of Wyoming and Montana.

While air pollution regulations represent the greatest single force acting on coal development in the near term, coal conversion into synthetic fuels may represent the greatest long-term influence on coal demand. Because of dwindling supplies of domestic oil and gas, the demand for substitute energy would be enormous if the relative costs of synfuels become competitive with domestic gas production. Various forecasts have placed synfuel output for the turn of the century at 2.7–13.0 quads. These rather sizable outputs are, of course, predicated on favorable economics and solutions to environmental and resource problems.

Though any estimates on the rate of development for the coal synfuels industry would be quite speculative, the probable locational patterns of such an industry are fairly evident. Coal and water availability will be primary factors in choosing sites, followed by proximity and access to markets. For these reasons, the Northern Great Plains, particularly North Dakota, represent an attractive area for synfuel conversion. Likewise, the Interior coal province may be the center for a great deal of synfuel conversion. Because of its enormous coal reserves, plentiful water, and proximity to large demand centers, Illinois may become the leading state for coal gasification.

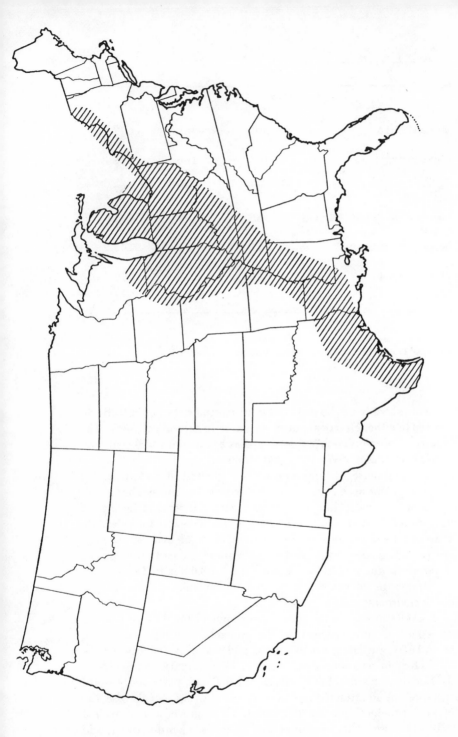

Figure 6.2   Coal Market Penetration Areas

DESCRIPTION OF ANALYSIS

The goal of most studies describing the socioeconomic impacts accompanying energy development has been to evaluate the expected effects on local employment, population, and community needs once a particular technology has been sited in a particular area. It is the goal of this study to demonstrate the utility of considering the socioeconomic effects prior to the selection of particular technologies for particular sites. This analysis will consider the sensitivities of these impacts to alternative types and levels of coal development and. to the economic, demographic, and geographic characteristics of different local areas. It will demonstrate that alternative siting patterns for various technologies can directly affect the extent of the associated socioeconomic impacts.

In what follows, the factors assumed to influence the rate and magnitude of socioeconomic impacts are identified and a framework for their evaluation is outlined. Because these impacts are assumed to vary with both the type and level of coal development as well as with the characteristics of the counties in which they are sited, a number of different coal technologies have been selected for study in a variety of counties from four coal provinces. The selection process for these "subject" counties and technologies is also described in this section.

Research Objectives

An important consequence of site-specific coal extraction and conversion activities is their potential effect on local (i.e., municipal and county) socioeconomic systems. Recent evidence of the nature of these socioeconomic consequences indicates that when the availability of jobs created by an energy facility exceeds the supply of willing local workers, it stimulates both in-migration and ancillary local economic activity. In some cases, especially where growth of available jobs is large relative to previously existing economic activity and population, economic, social, and political institutions have been unable to expand rapidly enough to accommodate increased needs for local services. Under such circumstances, when the expected living standards of both old and new residents are not satisfied, out-migration begins, productivity decline caused by labor turnover ensues, and a self-perpetuating cycle of social and economic impact is begun.

In other cases, however, the stimulus of new employment opportunities has an opposite effect. In areas subjected to development, which provides new jobs for unemployed or underemployed members of the local labor force, the social and economic effects may be positive. In such instances, the cycle of natural out-migration and economic decline may effectively be reversed by the advent of energy development activities.

Unfortunately, the nature and extent of social and economic effects is dependent upon a number of factors particular to both the type and size of the proposed facility and the economic, demographic, and geographic characteristics at alternative sites. For example, the rate and magnitude of socioeconomic changes is directly dependent upon:

- The capital and manpower requirements of the proposed energy facility
- The availability of local workers at the site
- The size of the labor pool within commuting distance of the site
- The accessibility of public and private services
- The adequacy and accessibility of available housing.

This dependence of the socioeconomic impacts on the site-specific nature of the energy facility and the area being developed makes generic forecasts of these impacts extremely difficult. If, however, the relative influence of each of these factors on prospective population growth and services requirements could be assessed, the socioeconomic consequences associated with alternative sites and various coal technologies could be quickly evaluated and consistently contrasted. It is precisely this type of sensitivity analysis that this report attempts.

Regions and Counties Considered

In order to examine the sensitivities of socioeconomic impacts to alternative economic, demographic, and geographic site characteristics, nine counties from four coal regions have been chosen for analysis. The first step in this selection process was to identify four states representative of four coal regions. The states and regions represented are:

- North Dakota (Fort Union Basin)
- Illinois (Interior coal province)
- Montana (Powder River Basin)
- West Virginia (Appalachian region).

The next step in the selection process was to choose coal resource counties from each of these states that were representative of the ranges of county sizes in each state. This step was to demonstrate the effect of predevelopment population size on the severity of socioeconomic impacts. Within each state, the counties were arrayed into population tertiles, and coal resource counties most nearly representative of the counties in each tertile were selected. Table 6.4 presents the array of county sizes by tertile for each state, the coal resource counties selected for analysis, their populations, and estimated reserves by types of coal. Notice that North Dakota has only one selected county and that West Virginia is restricted to

Cumulative Frequency %	Base Year Population	Coal Resource County	Base Year Population	Estimated Reserves in 10^6 Tons and Type of Coal
		Montana		
0%	0			
35%	4,000			7313 Strip Bituminous
		Rosebud	6,032	
70%	11,000			1150 Strip Bituminous
		Custer	12,174	
100%	90,000	Yellowstone	87,367	590 Deep Bituminous
		Illinois		
0%	0			2440 Deep Bituminous
		Hamilton	8,665	
33%	17,000			1425 Strip Bituminous
		Perry	19,757	
67%	38,500			975 Strip Bituminous
		Peoria	195,318	
100%	500,000+			
		North Dakota		
0%	0			
34%	4,800	Dunn	4,895	2000 Strip Lignite
67%	9,000			
100%	75,000			
		West Virginia		
0%	0			
33%	13,000			
		Wyoming	30,095	1640 Deep Bituminous
70%	35,000			
		Monongolia	63,714	3000 Deep Bituminous
100%	125,000-			

Source: *The Reserve Base of U.S. Goals by Sulfur Content (in 2 parts)*, Bureau of Mines Information Circular - 1975, IC-8693, U.S. Department of Interior.

Table 6.4   County Tertiles, Representative Counties, and Estimated Reserves

two counties. These restrictions were made to limit the analysis, because the selected counties are fairly representative of the counties in those states having considerable reserves. The geographic location of each of the selected counties is illustrated in Figure 6.3.

Alternative Coal Technologies Considered

In addition to the preimpact characteristics of the counties, local socioeconomic impacts are also sensitive to types and levels of coal development. In order to assess the sensitivities of socioeconomic impacts to alternative levels and types of coal development, four coal technologies have been selected for analysis. The four types of coal technologies and their associated levels are:

- 250 million cubic feet/day (Mcf/d) Lurgi coal gasification
- 800 megawatt (MW) coal electric generating facility
- 1600 MW coal electric generating facility
- 6 million (M) tons/year coal extraction activity.

It is the level of direct employment for each of these activities that is of primary importance in forecasting the associated socioeconomic impacts. However, the nine counties of the four coal regions contain both lignite and bituminous coal, which is either strip- or deep-mined. Therefore, the nature of the mining activity and the number of tons per year required as feedstock for each of the conversion facilities in each of the counties must be estimated before direct employment estimates are made. Using estimates obtained from the Bureau of Mines and checked against existing levels of mine employment recorded in the *Keystone Manual*, Table 6.5 presents the direct employment requirements by type and size of mine for each coal development activity in each of the nine counties.

The levels of employment at each of the conversion facilities are presented in Table 6.6. It has been assumed that the numbers of workers required to construct and operate these facilities will remain invariant across regional boundaries. As indicated, not all technologies are evaluated in each of the nine counties. It was determined that the sensitivities of alternative types and levels of coal development could be demonstrated without unnecessary duplication of analysis. Table 6.7 presents a summary of the types of technologies evaluated in each county.

Framework of Analysis

In order to evaluate the sensitivities of local socioeconomic impacts, which can range from positive (beneficial) to negative (detrimental) changes, estimates of the numbers of direct and indirect jobs created

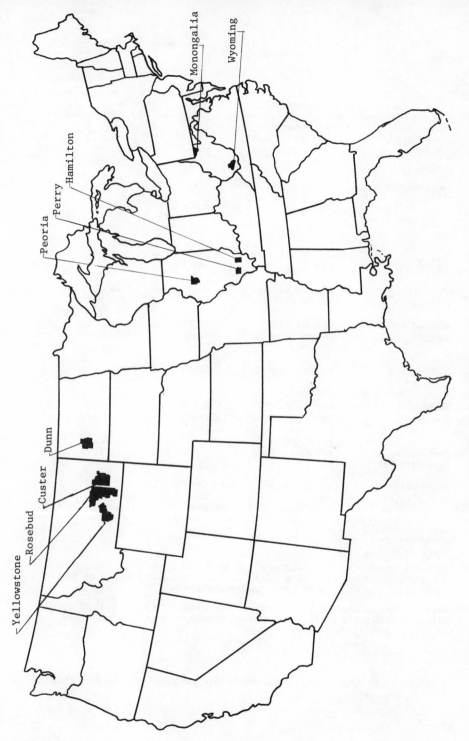

Figure 6.3　Map of Selected Coal Resource Counties

County State	Technology	Size of Mine	Type of Mine	Number of Employees Per Year*
Dunn County North Dakota	Gasification 1600 MW Electric 800 MW Electric 6 MM Mine	9MM 6MM 3MM 6MM	Strip Strip Strip Strip	200 100 75 100
Hamilton County Illinois	Gasification 800 MW Electric 6 MM Mine	6MM 2MM 6MM	Deep Deep Deep	1260 425 1260
Perry Co. Illinois	Gasification 1600 MW Electric 800 MW Electric 6 MM Mine	6MM 4MM 2MM 6MM	Strip Strip Strip Strip	180 150 100 180
Peoria Co. Illinois	1600 MW Electric 800 MW Electric	4MM 2MM	Strip Strip	150 100
Rosebud Co. Montana	Gasification 800 MW Electric 6 MM Mine	7.5MM 2.5MM 6MM	Strip Strip Strip	130 70 100
Custer Co. Montana	800 MW Electric	2.5MM	Strip	70
Yellowstone Co. Montana	Gasification 800 MW Electric 6 MM Mine	7.5MM 2.5MM 6MM	Strip Strip Strip	130 70 100
Wyoming Co. West Virginia	800 MW Electric	2MM	Deep	425
Monongalia West Virginia	Gasification 800 MW Electric 6 MM Mine	5MM 2MM 6MM	Deep Deep Deep	1100 425 1500

Sources:

Basic Estimated Capital Investment and Operating Costs for Underground Bituminous Coal Mines, U.S. Department of the Interior, Bureau of Mines Information Circulars 8689 (1975) and 8682A (1976).

Basic Estimated Capital Investment and Operating Costs for Coal Strip Mines, U.S. Department of the Interior, Bureau of Mines, Information Circulars 8661 (1975) and 8703 (1976).

Table 6.5   Direct Mining Employment by Type and Size of Mines for Each Technology in Each County

250 Mcf/d Lurgi Coal Gasification Employment Estimates, All Regions:

Year	Construction	Operation	Total
1	355		355
2	1270		1270
3	2070		2070
4	2320		2320
5	585		585
6		630	630
7		630	630
8		630	630
...			
10		630	630
...			
20		630	630
...			
30	.	630	630

Coal-Electric Generating Stations Employment Estimates, All Regions:

Year	400 MW Unit 1	400 MW Unit 2	400 MW Unit 3	400 MW Unit 4	800 MW Facility	1600 MW Facility
1	420				420	420
2	840	420			1260	1260
3	420	840	420		1260	1680
4	55	420	840	420	420	1080
5	55	55	420	840	110	1260
6	55	55	55	420	110	420
7	55	55	55	55	110	220
8	55	55	55	55	110	220
...						
10	55	55	55	55	110	220
...						
20	55	55	55	55	110	220

Sources: Woodward Envicon, Inc., *Socioeconomic Characterization and Assessment of the North Dakota Coal Gasification Project Area,* prepared for the Michigan-Wisconsin Pipeline Company, Sept. 1974.

Manpower, Materials, Equipment, and Utilities Required to Operate and Maintain Energy Facilities, Bechtel Corporation, March 1975, U.S. Dept. of Commerce, NTIS PB-255 438.

Dalstad, N.L., et al., *Economie Impacts of Alternative Energy Development Patterns in North Dakota,* Final Report, North Dakota State University, Fargo, North Dakota, June 1974.

18th Steam Station Cost Survey, Electrical World, Nov. 1, 1973.

Table 6.6   Direct Facility Employment Requirements by Type and Size of Facility

	Gasification With Mine	800 MW With Mine	1600 MW With Mine	6 M Ton/Yr Mine	800 MW Without Mine
1. Dunn	X	X	X	X	
2. Hamilton	X	X		X	
3. Perry	X	X	X	X	
4. Peoria		X	X		X
5. Rosebud	X	X		X	
6. Custer		X			
7. Yellowstone	X	X		X	
8. Wyoming		X			
9. Monongalia	X	X		X	

Table 6.7   Technologies Evaluated by County

annually, the annual changes in local employment, and the numbers and sizes of households in-migrating annually have been prepared for each of the 26 separate cases summarized in Table 6.7. In addition, estimates of the spatial distribution of in-migrating households among neighboring communities and the increased requirements for local services implied by these new residents have also been prepared for several of the more severely impacted areas.

A complete discussion of the methods used in forecasting these changes is presented in Appendix A. The results of the analyses for each of the 26 separate cases are presented in tabular fashion in Appendix B.

ASSESSMENT OF IMPACT SENSITIVITIES

In this section, the method of analysis just described is used to reveal the sensitivities of socioeconomic impacts to different types and levels of coal development in counties possessing different economic, demographic, and geographic characteristics.[2] This assessment will demonstrate that while interregional county differences are important, intraregional siting decisions are also important in reducing socioeconomic impacts. It will also be demonstrated that within similar counties, the type and level of coal development sited can materially alter the nature of the associated impacts, changing them from essentially positive to negative. Finally, an attempt will be made to isolate those county and technology characteristics that are of major importance in determining the levels of employment and population changes.

Adverse Impacts from Coal Development

The example presented in Appendix A of coal gasification in Dunn County illustrates the nature of the employment and population changes capable of precipitating adverse or negative socioeconomic effects. In that example, a definite "boom-bust" cycle' of population change was projected. This rapid relative change in population of the county was demonstrated to cause significant problems with respect to the growth and decline of local communities and the associated needs for major temporary changes in service facilities. However, not all of the projected associated impacts in Dunn County are negative. It was also revealed that 1,336 persons from Dunn and surrounding counties not otherwise employed would find permanent jobs at the gasification plant and in the local service sector. These increased employment opportunities may well reverse a 20-year trend in out-migration in western North Dakota.

Coal gasification in Dunn County is, however, neither an isolated nor an exceptionally severe example of a county confronting adverse impacts from coal development. Rosebud County, Montana, had a base population of 6,032 in 1970. It is representative of the smaller counties in Montana, lying just outside of the first population tertile. Although slightly larger than Dunn County, Rosebud can expect less local employment and more in-migration than Dunn County. Figure 6.4 illustrates the differences in population and employment impacts quite clearly. These differences are even more interesting in light of the fact that fewer basic direct jobs are associated with the gasification plant and mine in Rosebud County.

Rosebud County has a current ratio of total (basic plus secondary) to basic jobs of 2.2, which is slightly higher than the ratio of 1.6 for Dunn County. Thus, the economic characteristics of the two counties are somewhat different, Rosebud having to supply more services locally than Dunn. There are also major demographic and geographic differences. Rosebud is a large, sparsely populated county surrounded, for the most part, by other good-sized, thinly populated counties. This situation accounts for the fact that few otherwise unemployed persons are expected to be available and willing to commute to the site of the new gasification plant. Thus, a significant proportion of the new jobs in both the basic and secondary (indirect) sectors will have to be filled by workers from beyond commuting distances who, it is assumed, will take up residence in Rosebud County. Table 6.8 indicates the likely annual settlement patterns of these in-migrating households within the communities of Rosebud County. Because the gasification plant and mine are assumed to be located just beyond the borders of Colstrip, the towns of Castle Rock and Colstrip are projected to receive the major impacts of the increase in population. As Table 6.8 indicates, Colstrip is expected to receive 3,047 new residents by

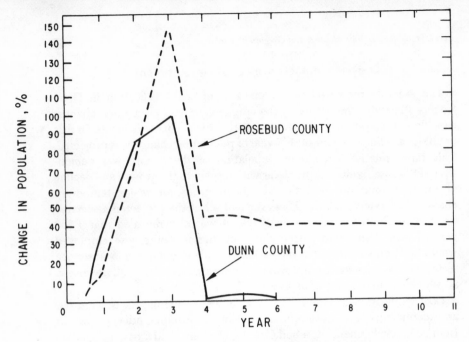

(a) Percentage Change in Total Population due to Immigration.

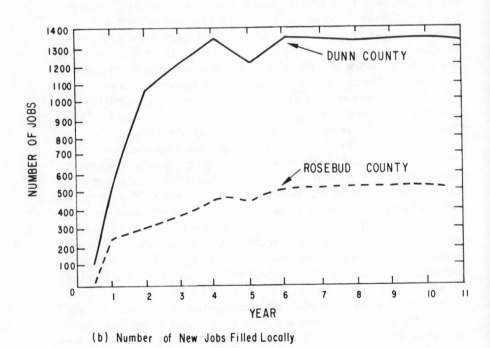

(b) Number of New Jobs Filled Locally.

Figure 6.4   Impacts from Coal Gasification in Dunn and Rosebud Counties

Town and Population / Year	Ashland 150		Castle Rock 100		Colstrip 100		Forsyth 1853		Lee 100		Rosebud 140	
	#	%	#	%	#	%	#	%	#	%	#	%
1	9	6	21	21	288	288	18	1	8	8	1	1
2	49	33	105	105	1469	469	91	5	41	41	4	3
3	86	57	186	186	2596	2596	161	9	72	72	7	5
4	102	68	219	219	3047	3047	189	10	85	85	8	6
5	30	20	66	66	927	927	57	3	26	26	2	1
6	31	21	66	66	927	927	57	3	26	26	2	1
7	32	21	73	73	1019	1019	63	3	28	28	3	2
8	32	21	73	73	1019	1019	63	3	28	28	3	2

Table 6.8   Spatial Allocation of New Households in Rosebud County by Year

the end of the fourth year. Castle Rock is expected to increase by 219 percent in the same period. The associated increased local service requirements for these two communities are presented in Tables 6.9 and 6.10. Table 6.10 indicates that up to 47 elementary classrooms will be required in the fourth "peak" year of new population in Colstrip. However, by the fifth year, only one-third of these classrooms will be required on a permanent basis.

The preceding analysis has served to demonstrate the nature of the relationship between local employment and population changes and community needs. In the following section, the impacts from alternative coal technologies within the same county are examined. Since service needs are directly correlated with relative population impacts, the succeeding assessments will concentrate on the timing and magnitude of population impacts as substitutes for the total socioeconomic effects of energy development.

Effects of Alternative Technologies in the Same County

In this section, the impacts of a single 6 M ton/year strip mine, an 800 MW electric generating station, and a gasification plant are assessed for Dunn County, North Dakota, and Yellowstone County, Montana. Yellowstone County, with a 1970 population of 87,367, ranked as one of the largest counties in Montana. Its economy, which supplies many of the services required by adjacent smaller counties, had an above-average concentration of indirect employment. The simple ratio multiplier in 1970 was 3.4. It has been assumed that the multiplier would decline to at least 3 with the introduction of a major coal-related activity. Although the county is surrounded by sparsely populated counties, there are a considerable number of persons living in Yellowstone who are not currently employed members of the labor force and who are able to accept jobs. This condition is due primarily to the fact that the female labor force participation rate is currently well below the national average.

Figure 6.5 depicts the change in population caused by in-migration and the number of new jobs filled by the local commuting work force for each of the three technologies. Although as many as 5,950 persons are expected to move into Yellowstone County in the third year of constructing the gasification plant, the relative impact on the total population of the county remains small. In fact, it is quite likely that this temporary increase in population can be absorbed by communities having sufficient excess capacities in their local infrastructures to avoid any strain on local services.[3] The fact that large numbers of local residents are finding jobs as a result of gasification recommends the siting of this type of facility in Yellowstone County.

Service Type	Year 1 (0)	Year 2 (92)	Year 3 (162)	Year 4 (190)	Year 5 (58)	Year 6 (58)
1. Elementary Classrooms	0		2-3 classrooms	3 classrooms		
2. High School Classrooms	0		1 classroom	1-2 classrooms		
3. Recreation Facilities	0		.6 acre playgrnd .5 acre park	.8 acre playgrnd .7 acre park		
4. Police Services	0		1 new officer	1 new officer		
5. Police Facilities	0		200 sq. ft. in station	200 sq. ft. in station		
6. Fire Services	0		none	none		
7. Fire Facilities	0		none	none		
8. Solid Waste Services	0		575 man-hrs.	660 man-hrs.		
9. Library Facilities	0		900 sq. ft. in library	1000 sq. ft. in library		

Table 6.9   Requirements of Selected Local Services for New Households in Castle Rock, Montana

Service Type	Year 1 (288)	Year 2 (1469)	Year 3 (2596)	Year 4 (3047)	Years 5-8 (1000)
1. Elementary Classrooms	5 classrooms	24 classrooms	42 classrooms	47 classrooms	16 classrooms
2. High School Classrooms	2 classrooms	10 classrooms	18 classrooms	20 classrooms	7 classrooms
3. Recreation Facilities	1 acre playgrnd .8 acre park	6 acre playgrnd 5 acre park	10 acre playgrnd 8.5 acre park	12 acre playgrnd 10 acre park	3.9 acre playgrnd 3.3 acre park
4. Police Services	1 new officer	4 new officers	7 new officers	9 new officers	3 new officers
5. Police Facilities	200 square feet in station	800 square feet in station	1400 square feet in station	1800 square feet in station	600 square feet in station
6. Fire Services	1 new fireman	3 new firemen	5 new firement	7 new firemen	2 new firemen
7. Fire Facilities	100 square feet in station	400 square feet in station	900 square feet in station	1080 square feet in station	360 square feet in station
8. Solid Waste Services	1000 man-hrs.	5200 man-hrs.	9035 man-hrs.	10,500 man-hrs.	3475 man-hrs.
9. Library Facilities	1700 square feet in library	9000 square feet in library	15,000 square ft. in library	18,000 square ft. in library	6000 square feet in library

Table 6.10 Requirements of Selected Local Services for New Households in Colstrip, Montana

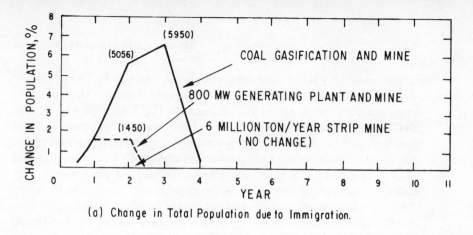

(a) Change in Total Population due to Immigration.

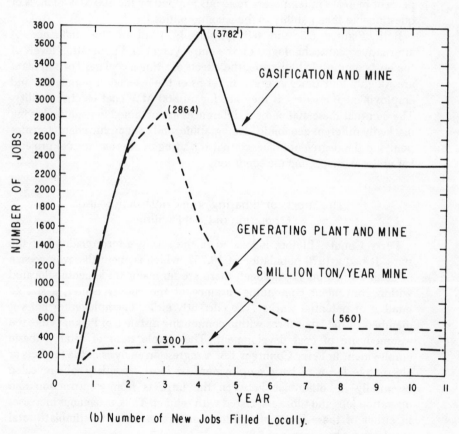

(b) Number of New Jobs Filled Locally.

Figure 6.5 Impacts Associated with Mining, Electric Generation, and Gasification in Yellowstone County, Montana

The rate of employment and population change in Dunn County associated with the three technologies is far more dramatic. The changes in population and employment are presented in Fig. 6.6. As part (a) reveals, the choice of the technology sited in Dunn County will have a direct bearing upon the extent of employment and population generated within the county. While both the gasification and 1600 MW electric facilities and associated mines result in considerable additional local employment, they carry a concomitant burden of major population changes. As shown in Table 6.A.5, such levels of in- and out-migration create definite problems for the communities expected to absorb the new residents. Even the 30 percent increase in temporary residents implied by the 800 MW plant is of questionable acceptability to these communities.

Both Figures 6.5 and 6.6 graphically illustrate the influence of alternative coal technologies on the employment and population levels of the host counties. In both cases, the effects of siting a coal gasification plant are roughly four times as great, in terms of the associated population and employment increases, as they are for an 800 MW coal electric facility. The fact that these 400 percent differences are obtained in counties having markedly different economic, demographic, and geographic characteristics points to the degree of influence that the siting of alternative coal projects has on local socioeconomic conditions.

The Effects of "Phasing" the Construction and Operation of Coal Facilities

Perry County, Illinois, is located in the southwestern quadrant of the state. It had a 1970 population of 19,757, which is about the average size of counties in Illinois. Although there are no major trade centers located within convenient commuting distance of the county center, a sizable number of potential workers are currently either unemployed or not yet members of the labor force within commuting distance of Pickneyville, the assumed site of coal development. The simple ratio of total to basic employment in Perry County is 1.8. A regression analysis of the impacts of alternative forms of basic employment on local secondary jobs revealed essentially no difference between the impacts from construction and operation jobs and those associated with mining. Thus, exogenous increases in either of these forms of employment are assumed to stimulate total employment by a factor of 1.8.

The differences expected to be caused in local population and employment by the phased construction and operation of a 1600 MW coal electric station as opposed to the simultaneous construction of four MW units are indicated in Fig. 6.7. As is evident from this figure, there are major differences in the impacts associated with these two levels of development.

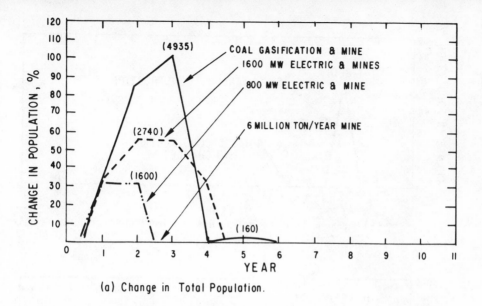

(a) Change in Total Population.

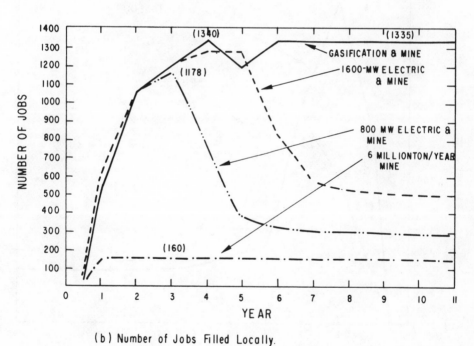

(b) Number of Jobs Filled Locally.

Figure 6.6   Impacts Associated with Mining, Electric Generation, and Gasification in Dunn County, North Dakota

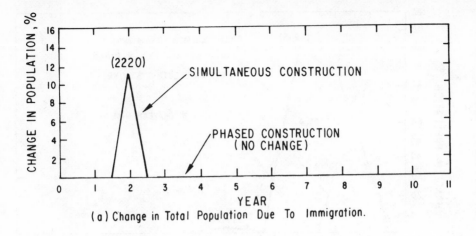

(a) Change in Total Population Due To Immigration.

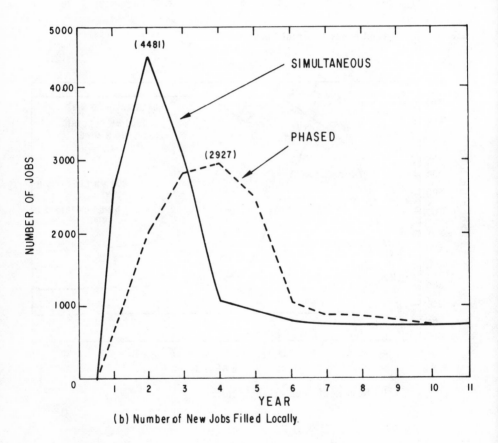

(b) Number of New Jobs Filled Locally.

Figure 6.7   Impacts Associated with Simultaneous and Phased Construction and Operation of a 1600 MW Coal Electric Generating Station

These differences are primarily caused by the differences in the number of construction workers required in years one through four. The construction work force figures for these two levels of development have been taken from the estimates presented in Table 6.6 and are repeated here.

Construction Work Force

Year	Phased	Simultaneous
1	420	1680
2	1260	3360
3	1680	1680
4	1680	0
5	1260	0

The extraordinary construction requirements involved in simultaneously building four 400 MW units includes the need for more than the number of available workers in Perry and adjacent counties. The result is that an estimated 1,000 jobs will have to be filled by in-migrating workers in the second year. This in-migrating work force is expected to increase the population of Perry County by 2,220 residents in this period. On the other hand, phasing the construction of these units permits local workers to fill almost all of the available jobs, resulting in a negligible increase in the population of Perry County.[4] In addition, phasing the construction of these units reduces the "boom-bust" nature of local employment. Jobs are distributed more evenly over time, which tends to moderate and prolong the associated local economic expansion. The resulting expansion in economic activity will encourage the retention and employment of a greater number of local people for a longer period of time. Thus, the economic expansion accompanying phased energy development in Perry County will be, for the most part, shared among the permanent residents of the area.

The Effects of Transporting Coal for Use in Other Regions

As illustrated in Figure 6.4, the effects of mining and gasifying coal in Dunn County, North Dakota, were expected to be quite severe. As a result of the presence of a single 250 Mcfd gasification plant and associated mine, the population of the county was projected to increase by the second and third years to a level double its current size. The effects of this rate and level of population growth were found to necessitate rapid and significant changes in the local public service infrastructure in Killdeer, North Dakota (see Table 6.A.5). However, as illustrated in Figure 6.6, the effects of opening and operating a 6 M ton/year strip mine to recover low-sulfur lignite coal in Dunn County were expected to be strictly positive. In this instance, no in-migrating workers were expected, while 160 new jobs were projected to be created for an expanded labor force.

Gasification of coal in Rosebud County would have even more impact than in Dunn County. This conclusion is graphically illustrated in Figure 6.4. The associated community population and service impacts among the towns in Rosebud County are presented in Tables 6.8, 6.9, and 6.10. Although rich in low-sulfur coal reserves, both of these small Northern Great Plains counties are unable to bear population and employment impacts associated with single 250 Mcfd gasification facilities without undergoing significant shortages in their public and private sectors.

Perry County, Illinois, on the other hand, can be expected to obtain a net socioeconomic benefit from a single 250 Mcfd gasification plant. As explained previously, Perry County ranks in the second population tertile in Illinois and has a significant projected available work force. Figure 6.8 illustrates the population and local employment effects of a coal gasification facility located in Perry County. Although this example contains the figures for mining employment and thereby slightly overstates local employment opportunities associated with a gasification facility, the point is that areas similar to Perry County can expect to experience the following net socioeconomic benefits from coal gasification:

• Expanded local employment and income
• Increased basic economic activity
• Retention of a share of local residents who might otherwise leave the area.

This example demonstrates that mining low-sulfur Western coal in small northern Great Plains counties and converting it in counties like Perry can result in positive socioeconomic impacts in each location. Clearly, the separability of extraction and conversion activities represents a viable tool for the migration of interregional socioeconomic impacts.

Effects of Similar Levels of Coal Development Among Similar Counties in Different Regions

The counties used in this analysis that are in the first (smallest) population tertiles of their respective states includes: Rosebud, Montana; Dunn, North Dakota; and Hamilton, Illinois. The population and employment impacts associated with an 800 MW electric generating station and associated mine in each of these counties are presented in Figure 6.9. This comparison is made to illustrate the effects of regional differences among similarly ranked counties on the employment and population changes caused by development. The most dramatic difference among these similarly ranked counties, made apparent in Figure 6.9, is the availability of a local supply of labor to fill new jobs.

Hamilton County, although of the same relative size as Rosebud and

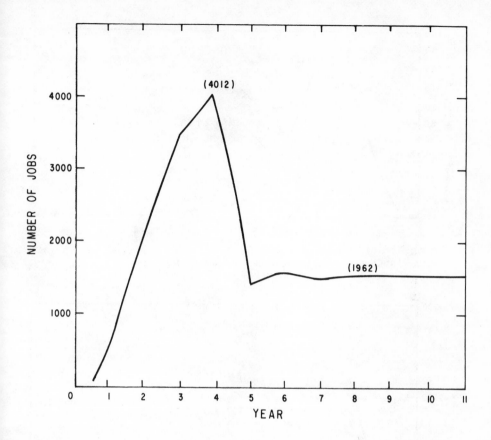

Figure 6.8   Impacts from Local Gasification in Perry County, Illinois

Dunn, is located in a region of greater population density, economic activity, and diversification. The higher availability of new local entrants to the labor force in this region is clearly a significant factor for the projected impacts of the 800 MW facility on population changes. The variability in the employment multipliers among these counties contributes further to these projected differences. Rosebud County has a current ratio multiplier of 2.2, while Dunn County's multiplier is only 1.6. This difference indicates that twice as many secondary workers are required to support additions to the basic work force in Rosebud as in Dunn County. Thus, even if the available supply of local workers were equal in Dunn and Rosebud counties, the total impacts on employment and population would be greater in Rosebud County.

These differences in the ratios of total to basic employment are not,

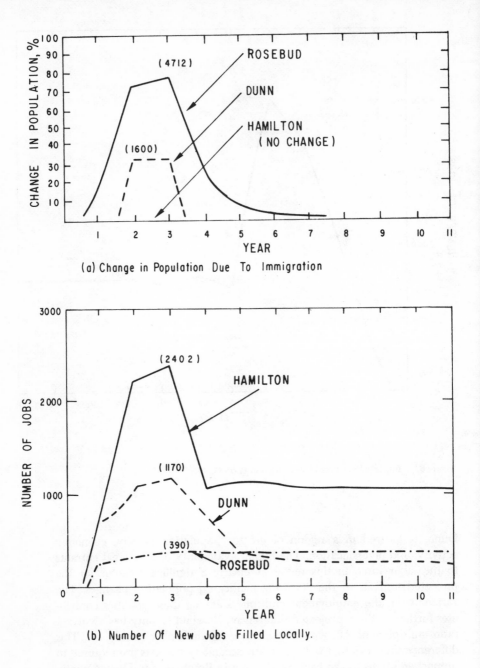

Figure 6.9 Impacts from 800 MW Electric Generation Facilities in Rosebud, Dunn, and Hamilton Counties

however, entirely a function of regional differences. That is, such differences are also influenced by intraregional geographic and economic factors. The multiplier is influenced by the nature of the major economic activity in the county and the relative accessibility of major trading centers. In Yellowstone County, Montana, the multiplier is currently 3.0. This relatively high ratio of total to basic jobs is largely attributable to the importance of tourism in the area and the fact that numerous secondary establishments exist to service this industry. It is also partly explained by the fact that Billings is a major trade center for a number of surrounding counties and has, therefore, a relatively large proportion of retail, commercial, and service industries, which draw their support from the basic ranching and farming industries.

The effects of regional differences among similar counties experiencing similar types of energy development can also be examined for counties having midrange populations for their respective states. In this analysis, the following counties are representative: Custer, Montana (12,174); Perry, Illinois (19,757); and Wyoming, West Virginia (30,095). The employment and population impacts associated with an 800 MW coal electric generating facility have been forecasted for these counties. The results are presented in Figure 6.10. As in the previous example, the representative county for Montana is expected to be most severely impacted. Thus, on the basis of the evidence presented in Figure 6.10, it would appear, ceteris paribus, that the average-sized counties of West Virginia and Illinois are better able to avoid negative socioeconomic impacts from the construction and operation of 800 MW coal-fired power plants than a medium-sized county in Montana. In fact, the increases in new jobs forecasted for these areas lend support to the conclusion that this form of energy development may result in positive economic impacts in Perry and Wyoming counties.

The regional differences in employment and population impacts associated with 800 MW coal-fired power plants have also been examined for counties representative of the largest population tertiles in Montana, Illinois, and West Virginia. The counties and their populations are: Yellowstone, Montana (87,367); Peoria, Illinois (195,318); and Monongalia, West Virginia (63,714). Figure 6.11 illustrates the nature of the annual employment and population projections for each of these counties. Again, Montana contains the only area expected to be impacted by in-migration as a result of the power plant and mine—Yellowstone in the Powder River Basin. Even Monongalia County, which is somewhat smaller than Yellowstone County in total population and must employ a greater number of workers to deep-mine the coal that the power plant requires, is not expected to experience an influx of new households. Both the employment multipliers and technologies in Yellowstone and Monongalia

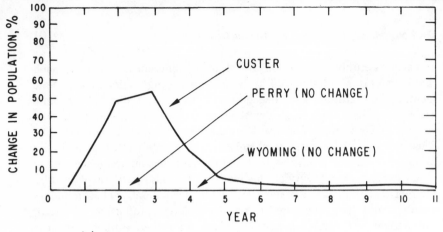

(a) Change in Total Population due to Immigration.

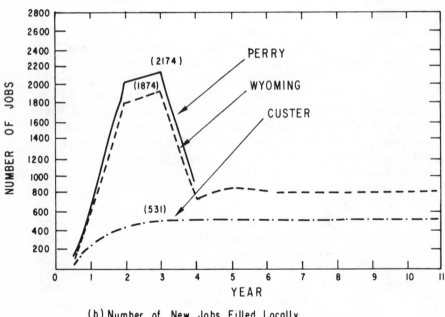

(b) Number of New Jobs Filled Locally.

Figure 6.10 Impacts from 800 MW Electric Generation Facilities in Custer, Perry, and Wyoming Counties

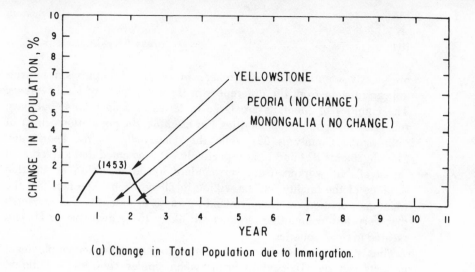

(a) Change in Total Population due to Immigration.

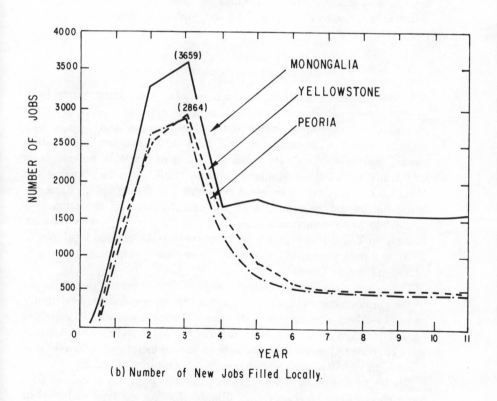

(b) Number of New Jobs Filled Locally.

Figure 6.11   Impacts from 800 MW Electric Generating Facilities in Yellowstone, Peoria, and Monongalia Counties

are nearly identical; the differences in projected population impacts appear to stem from the differences in the availability of local workers. The counties of Montana are large in terms of square miles but have relatively low population densities. For example, the population density of Yellowstone County is 33, while the density figures for Peoria and Monongalia are 314 and 175, respectively.[5] It is assumed that a fraction of the available workers from surrounding localities within commuting distance of the facility will be willing to fill new jobs. The fact that the Montana counties are characterized by these low density figures means that very few commuting workers are likely to be available to fill jobs created in these counties.

Thus, it appears that definite differences in population and employment impacts can be expected to occur when similar counties in different regions are exposed to similar rates and levels of development. The next section will examine the sensitivities of these impacts to intraregional differences in county economic and demographic characteristics.

Effects of Intraregional Differences on Employment and Population Impacts

In the preceding analysis, Montana counties were observed to have greater projected population impacts and, concomitantly, less local employment than their counterparts in the Interior and Appalachian regions. However, as indicated in Figure 6.12, considerable variation among the impacts associated with the siting of 800 MW power plants and mines within these Montana counties is also expected. Among the Montana counties having different economic and demographic characteristics, the impacts associated with the construction and operation of a single 800 MW facility and mine vary from expected peak population changes in Yellowstone County of 1.6 percent and increased local jobs of 2800, to a peak population increase of 78 percent and increased local jobs of 390 in Rosebud County.

The primary reasons accounting for these differences are the variations in the employment multipliers (economic differences) and the availability of local workers (demographic differences). Yellowstone, with its expected annual additions to the local labor force of 640 and a multiplier of 3.0, is in stark contrast to Rosebud County with an annual increment to its available labor supply of 80 persons and an employment multiplier of 2.2.

Similar intraregional differences in projected population and employment changes can be observed in Illinois. Holding the type and level of technology constant by observing the impacts associated with a single coal gasification facility and mine in Hamilton and Perry counties, Figure 6.13 illustrates the variations in impacts that are accounted for by differences in

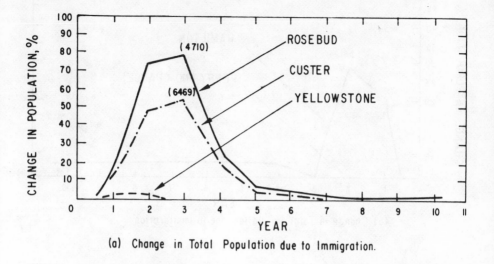

(a) Change in Total Population due to Immigration.

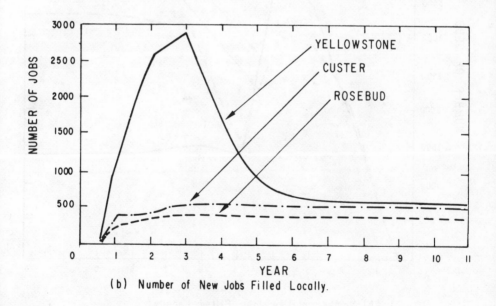

(b) Number of New Jobs Filled Locally.

Figure 6.12   Impacts from 800 MW Electric Generating Facilities in Rosebud, Custer, and Yellowstone Counties

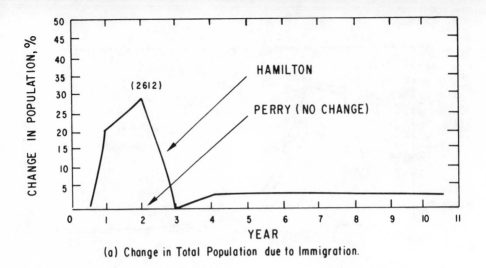

(a) Change in Total Population due to Immigration.

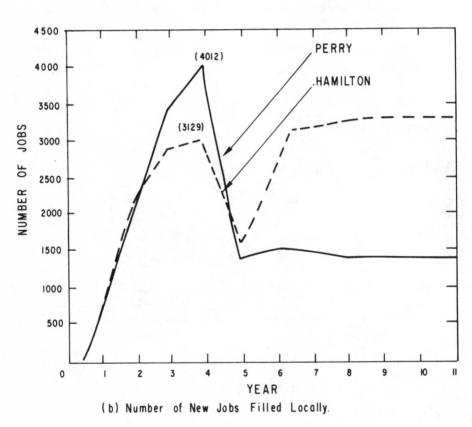

(b) Number of New Jobs Filled Locally.

Figure 6.13   Impacts from Coal Gasification in Hamilton and Perry Counties

the economic and demographic characteristics of these Illinois counties. As in the preceding Montana example, the economic and demographic differences between these two Illinois counties are expected to account for significant variations in the population and employment impacts. Hamilton County has a population of 8665, with a density per square mile of 20, while Perry has a population of 12,157, and twice the density. The projected annual increment to the local supply of labor in Hamilton and Perry counties is 290 and 385, respectively. However, estimates of unemployment and shortfalls in the local labor force participation rates by age and sex cohorts in these counties suggest that during the first year in the life of the new plant as many as 4000 new workers may be available in Perry County and only 2280 in Hamilton. There are no significant multiplier differences between these counties. Finally, it must be pointed out that the types of mines needed to supply the gasification facilities in these counties are different. Hamilton County requires several deep mines to support its plant, while the feedstock in Perry County is expected to come from a single strip mine. The employment differences between these two types of mines are reflected in the greatly expanded local job opportunities in Hamilton County in the sixth year.

SUMMARY AND CONCLUSIONS

This brief analysis has attempted to look systematically at the nature of the impacts associated with coal development and to demonstrate their sensitivities to differences in (1) the types and levels of development proposed, and (2) the economic, demographic, and geographic characteristics of the counties in which the development is sited. The purpose of this analysis has been to focus attention on the variability of socioeconomic impacts to these factors, and to provide evidence for public and private energy officials that, if careful consideration is given to these factors before facilities are sited, the impacts can be measurably reduced—and in some instances changed from negative to positive.

It has been demonstrated that the employment and population changes associated with energy development can have dramatic effects on the recipient communities. Earlier in this paper, projections of the annual changes in the population caused by gasification development in Rosebud County were computed. This population was then allocated among the localities of the county and the effects on the provisions of local services were estimated. It was concluded that the "boom-bust" cycle of population change and the dramatic shortfall in services required to preserve existing living standards would impose considerable economic and social costs on all of the parties-at-interest, including the local residents and the energy industry.

The adverse socioeconomic impacts associated with projected employment and population changes having been made obvious, an attempt was made to demonstrate the influence that alternative types and levels of energy development would have on county employment and population impacts. The results of this analysis in Dunn County, North Dakota, and Yellowstone County, Montana, revealed that in both counties, the population and employment impacts of coal gasification could be approximately four times as large as the impacts associated with an 800 MW electric generating facility. The fact that similar large differences in impacts were observed in these two diverse counties led to the conclusion that the type of technology has a significant independent effect on socioeconomic impacts.

The influence of the rate of development on employment and population changes was studied separately by examining the impacts associated with "phased" and simultaneous construction of a 1600 MW electric generating facility in Perry County, Illinois. The results of the analysis indicated that when properly phased, construction of these facilities can coincide with the availability of local labor and diminish the likelihood of adverse impacts. The Perry County example provided dramatic evidence of the positive effects that a more gradual construction schedule could have on the host county.

The effects of geographically separating the extraction and conversion activities were assessed. In the example (which it is stressed, was purely hypothetical) the impacts of gasifying low-sulfur coal from Dunn County, North Dakota, and Rosebud County, Montana, in Perry County, Illinois, were examined. It was observed that this arrangement made possible positive projected employment and population impacts in all three counties. This result stands in contrast to the previous examples of performing both coal mining and gasifying activities in these two small Western counties.

The preceding examples provide evidence that varying the type and intensity of energy development can directly influence the rate and magnitude of the associated socioeconomic impacts. These results also suggest that the choice of technologies and the selection of sites can be effective in mitigating these impacts, if consideration is given to these factors before a final decision is made on the location for a particular facility. Even if only one site exists that satisfies the criteria of resource and water availability, the analysis of this section suggests that phasing the construction of the facility can measurably alter the population and employment impacts.

The extent to which modifications of a given technology are needed to mitigate negative socioeconomic impacts is a function of the nature of the county being impacted. For example, the construction and operation of a

250 Mcfd gasification plant and associated coal mine in Perry County, Illinois, is expected to create fewer jobs than the projected local work force can accommodate. As a result, no negative impacts from untoward population growth are anticipated. However, a similar development in Rosebud County, Montana, is expected to overwhelm the local work force and result in a population increase of over 100 percent in under four years. These dramatic differences in expected results are clearly not attributable to differences in the technologies. Instead, they must be accounted for by differences in the characteristics of the counties themselves.

Although not exhaustive, a listing of the county characteristics that influence the extent of the impacts from development includes:

- The relative size and diversification of the existing county economic base
- Levels of employment and labor force participation
- The population density
- The accessibility of major trade centers
- The nature of the coal mining activity in the county (whether deep or strip)
- The nature of these characteristics in adjacent counties.

The influence of these economic, demographic, and geographic characteristics on the expected impacts from coal-related energy development were evaluated by projecting the employment and population changes in a variety of counties subjected to identical types and levels of development.

In the first series of cases, the variations in expected population and employment impacts were forecasted for counties of similar sizes located in different states and coal provinces. In the first example, the impacts from a single 800 MW electric generating facility and mine were projected for the three subject counties in the smallest population tertiles of their respective states. The results confirmed that county characteristics influence impacts and that there are definite differences among counties of different regions. It was found that the projected in-migration of basic and secondary workers would cause population increases of 4700 in Rosebud County, 1600 in Dunn County, and result in no new residents in Hamilton County. These differences in population impacts were found to be directly attributable to differences among these counties in their basic economic activities (multipliers), population densities, and the characteristics of adjacent counties.

Similar comparative analyses were conducted for the medium and large counties in Montana, Illinois, and West Virginia. In the analysis of Custer, Perry, and Wyoming counties, the impacts associated with an 800 MW facility and mine were projected to be negative for Custer County, Montana, but positive in both Perry County, Illinois, and Wyoming

County, West Virginia. Only among the larger counties of these states was there sufficient "carrying capacity" to permit the projected impacts from a single 800 MW facility and mine to be positive in all cases.

Although clearly not conclusive, the evidence of this series of analyses indicates that there appear to be fairly consistent regional differences in the characteristics of these counties, which account for the variations in impacts. For example, the Montana counties were observed to have greater projected population impacts and, concomitantly, less local employment than counties of relatively similar size in the Interior and Appalachian regions. However, variations in the economic, demographic, and geographic characteristics also were found among counties in the same state.

In a separate series of analyses, the effects of intraregional differences in county characteristics on socioeconomic impacts were evaluated by again holding the type and level of energy development constant. The effects were examined separately for the counties of Montana and the counties of Illinois. Although the population size differences in these counties appeared to account for a considerable proportion of the differences in the projected changes in employment and in-migration, it was interesting to observe the differences among these counties in terms of other characteristics. For example, the employment multipliers, which reflect the diversity of the economy and the proportion of basic economic activity, ranged from 2.2 to 3.0 in Montana, and from 1.8 to 2.4 in Illinois. In addition, population densities varied from one to 33 persons per square mile in the Montana counties and from 20 to 314 persons per square mile in the Illinois counties. Thus, while there appear to be regional differences in the economic, demographic, and geographic characteristics of coal-resource counties, significant variations in these characteristics can also be found among counties of the same state.

This study has examined the effects on projected socioeconomic impacts of alternative types and levels of coal-related energy development and differences in the preimpact economic, demographic, and geographic characteristics of coal-resource counties. The results illustrate the relationship between site selection and associated socioeconomic impacts. They indicate that identical technologies can have the effects of expanding local employment opportunities and reversing out-migration in one area while precipitating a process of "boom-bust" population growth and decline in another.

While the results of the analyses do not prove conclusively which counties and/or regions will benefit from energy development, they do indicate that medium- and small-sized counties in the Northern Great Plains appear less likely to benefit from coal development than do their counterparts in Illinois and West Virginia. There is also evidence in this study to support a preference for large rather than small counties in the site-selection process. However, the considerable variations found in the economic, demographic, and geographic characteristics among counties both interregionally and intraregionally prohibits making further generalizations from these results.

This study has also demonstrated the effects that variations in the types and levels of development can have on the socioeconomic impacts in the same or similar counties. For example, the analysis suggests that if, for other reasons, a particular technology must be sited in a county that lacks the "carrying capacity" to support the anticipated changes, the negative impacts can be reduced by "phasing" the construction of the facility.

Finally, and most importantly, the dramatic variations in the projected impacts associated with different technologies in different counties clearly illustrate the need to consider the socioeconomic impacts before rather than after plant siting decisions are made.

APPENDIX A.   Forecasting Methods

The employment figures for each technology represent only a portion of the new jobs created as a result of energy development. The new direct energy-related jobs represent additions to the basic employment of the area. Basic jobs, because they arise from and are paid for by sources beyond the impact area, stimulate the expansion of demand for and employment in the local secondary sector. Thus, the addition of basic economic activity will give rise to expansion in local secondary activity.

The relationship between basic and total jobs has been estimated in a variety of ways. Perhaps the simplest and most common procedure for estimating the effects on total employment of exogenous increases in basic jobs is through the use of the simple ratio multiplier. The ratio multiplier is calculated by summing the number of total jobs in the county (or area being studied) and dividing by the aggregate number of jobs that can be considered basic. Assuming that the existing relationship between total and basic jobs will be maintained in the future, this simple ratio provides a multiplier that, when multiplied by the number of basic jobs expected in any given period, yields an estimate of the concomitant increase in total employment.

There are, however, two problems associated with the use of this procedure. First, the simple ratio multiplier requires its user to assume that additions of any type of basic employment (e.g., agriculture, construction, government) will be equally stimulative. Secondly, the simple ratio multiplier merely provides an estimate of the equilibrium or final level of total employment. This multiplier procedure provides no indication of the timing of the creation of the new secondary jobs. Explicit consideration has been given to both of these estimation problems in the forecasts prepared for this paper.

The effects of exogenous increases (or decreases) in different types of basic employment have been empirically estimated for the counties in each region. Using cross-sectional employment data on similar counties, a regression equation was specified to reveal the differential effects on secondary employment of exogenous changes in alternative forms of basic employment. The results of this analysis, within each of the regions studied, revealed no significantly different impacts for construction, operations, or mining employment. Therefore, the simple ratio multiplier was assumed to apply equally to each of these types of basic employment.

Explicit analyses of the lag problem in the adjustment of secondary employment to immediate increases in basic employment were also undertaken. It was determined that employment changes in the secondary sector would generally arise from the changes in demands for local goods and services generated by the increased basic economic activity. Based on the assumptions (1) that a portion of each new dollar from the basic activity would be spent locally and (2) that the income velocity of money is approximately three, it was found that it would take approximately four years for secondary, and therefore total, employment to reach its new equilibrium after a single period change in basic employment.

Table 6.A.1 indicates that 355 new basic jobs are expected in the first year. Using the simple ratio multiplier, it is estimated that ultimately 210 secondary jobs will be created to provide for the needs of this increase in basic activity.[6] The lag model described above indicates that approximately 71 percent of the total increase in secondary jobs will be created in the first year, 17 percent in the second, 8 percent in the third, and the final 4 percent created in the fourth period. The corresponding secondary employment figures are presented in the first four rows of column three. Similar calculations are made for each of the six years in which changes in basic employment are expected to occur. The annual changes in total secondary employment are then obtained by summing across columns three through eight. The changes for each period are summarized in column nine.

Given the estimates of the annual changes in the number of total jobs for the county (column ten, Table 6.A.1), the impact on local population must be estimated. The first step in this process is to determine the quantity of available (not currently employed) workers willing to commute to the site of the increased employment activity. This step was done by projecting forward the available annual additions to the local labor force in each county in the continental United States.[7] The projections of the additional work force for the subject and surrounding counties were reduced by the

Year	Annual Change in Basic Employment	1 210	2 550	3 480	4 150	5 -1040	6 150	Annual Change in Secondary Employment	Annual Changes in Total Employment
1	355	150						150	505
2	915	35	390					425	1340
3	800	18	90	380				448	1248
4	250	10	45	80	105			240	490
5	-1735		23	40	25	-740		-650	-2385
6	245			20	12	-175	105	-38	207
7	0				6	-80	25	-49	-49
8	0					-40	12	-28	-28
9	0						6	0	6
10	0						0	0	0

Table 6.A.1 Employment Lag Estimates in Dunn County, North Dakota (Multiplier Equals 1.6 Simple Change in Secondary Employment [.6 * Annual Basic Changes])

percentage assumed unwilling to commute to the new jobs. The final estimates of the annual additions to the work force willing to commute to the site of the energy activity in each of the nine counties are presented in Table 6.A.2.

Once the total changes in the number of jobs and available local workers have been computed, the next step is to derive the number of jobs that cannot be filled by the local or commuting work force. These figures are presented in Table 6.A.3 for Dunn County, North Dakota. This table indicates that in the first year, 950 local workers are expected to be available to fill 505 new jobs. In the second year, an additional 1340 new jobs will be created but only 565 workers (the 445 remaining workers from the first year plus 120 new workers available in the second year) are available to fill them. Thus, 775 jobs will have to be filled by workers and their families in-migrating to Dunn County. In the third year, an additional 1115 jobs will have to be filled by an in-migrating work force. In all, there are a total of five periods in which in-migrating workers will

Table 6.A.2　Projections of Available Work Force in Impacted Counties

Dunn County, North Dakota

Year	Dunn County	Mercer County	Mc-Kenzie	Stark Co.	Billings County
1	310	200	200	200	40
2	40	15	15	50	2
3	60	10	15	50	2
4	60	10	15	50	2
5	60	10	15	50	2
6	60	10	15	50	3
7	50	10	15	50	3
8	40	10	15	50	3
9	40	10	15	50	3
10	40	10	15	50	3

Custer County, Montana

Year	Custer County	Rosebud County	Powder River Co.	Fallon County	Prairie County
1	250	30	10	40	10
2	75	8	2	5	3
3	75	10	2	8	3
4	75	10	2	8	3
5	75	10	2	8	3
6	75	10	2	8	3
7	70	10	3	8	2
8	70	10	3	7	2
9	70	10	3	7	2
10	70	10	3	7	2

Rosebud County, Montana

Year	Rosebud County	Treasure County	Mussel-Shell Co.	Custer County	Garfield County
1	136	15	15	70	3
2	40	3	3	25	1
3	55	3	3	25	1
4	55	3	3	25	1
5	55	3	3	25	1
6	55	3	3	25	1
7	55	3	3	25	1
8	50	3	3	25	1
9	50	3	3	25	1
10	50	3	3	25	1

Table 6.A.2   Projections of Available Work Force in Impacted Counties (cont'd.)

Yellowstone County, Montana

Year	Yellow-stone Co.	Big Horn County	Carbon County	Still-water	Mussel-Shell Co.	Treasure County
1	1600	120	60	60	35	3
2	600	15	8	5	5	1
3	700	15	8	6	5	1
4	700	15	8	6	5	1
5	700	20	8	6	5	1
6	700	20	8	6	5	1
7	600	20	8	6	4	1
8	600	20	9	6	4	1
9	600	20	9	5	4	1
10	600	20	9	5	4	1

Hamilton County, Illinois

Year	Frank-lin Co.	Jeffer-son Co.	Wayne County	White County	Gallat-in Co.	Saline County
1	834	286	91	123	170	398
2	70	47	21	24	12	36
3	83	52	27	30	14	41
4	83	52	27	30	14	41
5	84	52	27	30	14	41
6	83	51	27	30	14	41
7	80	49	26	28	12	41
8	68	48	23	24	13	38
9	68	48	23	24	13	38
10	68	48	23	24	13	38

Perry County, Illinois

Year	Jeffer-son Co.	Frank-lin Co.	Jack-son	Randolph	Washing-ton Co.
1	286	625	1760	375	176
2	47	52	211	40	22
3	52	62	76	47	31
4	52	62	76	47	31
5	52	62	76	47	31
6	51	62	76	47	31
7	49	60	74	47	29
8	48	51	77	47	31
9	48	51	77	47	31
10	48	51	77	47	31

Table 6.A.2   Projections of Available Work Force in Impacted Counties (cont'd.)

Peoria County, Illinois		Wyoming County, West Virginia		Monongalia County, West Virginia	
Year	Peoria County	Year	Wyoming County	Year	Monongalia
1	4008	1	3722	1	6353
2	473	2	261	2	703
3	385	3	290	3	344
4	385	4	290	4	343
5	371	5	290	5	343
6	364	6	290	6	343
7	364	7	280	7	337
8	364	8	224	8	301
9	364	9	˙224	9	301
10	364	10	224	10	307

be required to fill available jobs. In these five periods, Dunn County is expected to experience the impacts caused by an influx of new residents.

Survey data indicate that the average household size of in-migrating energy workers is 2.3 and that the average size of households in Dunn County is 2.9. Assuming that the in-migrating secondary workers have households similar in size to those of Dunn County residents, and that roughly 40 percent of the new in-migrating workers will fill secondary jobs, the weighted average household size of new workers will be 2.5. If, as census figures indicate, there are 1.2 workers per household, each new job that cannot be filled by a local worker will result in an expansion in population of 2.1 persons. Table 6.A.3 indicates that Dunn County will experience a total population increase of 4935 persons by the third year in the life of a coal gasification facility. This number is made the more significant when it is realized that it represents an increase of 100 percent in the total size of Dunn County.

Because this is a sizable impact in terms of new residents, it is important to examine the effects that this expanded population will have on the surrounding communities in which the newcomers are likely to locate. Using a simple gravity model of the form:

$$L_i = B \, D \, W_i \, d_{ij}^{-\alpha}$$

where

D = the number of new households

B = $1/\sum\limits_{i} W_i d_i^{-\alpha}$

Total New Jobs		Maximum Available Workers		Total Local Employment		Total Non-Local Employment		New Population*		New Households**	
Annual Change	Cumm. Total	Annual Change	Cumm. Total	Annual Change	Cumm. Total	Annual Change	Cumm. Total	Annual Change	Cumm. Total	Annual Change	Cumm. Total
505	505	950	950	505	505	0	0	0	0	0	0
1340	1845	120	1070	505	1070	775	775	1705	1705	645	645
1245	3095	135	1250	135	1205	1115	1890	2450	4155	930	1575
490	3585	135	1340	135	1340	355	2245	780	4935	295	1870
-2385	1200	135	1475	-140	1200	-2245	0	-4935	0	-1870	0
207	1407	135	1610	135	1335	72	72	160	106	60	60
-49	1358	125	1860	0	1335	-49	23	-110	50	-40	20
-28	1330	115	1975	-5	1330	-23	0	-50	0	-20	0
6	1336	115	2090	6	1336	0	0	0	0	0	0
0	1336	115	2250	0	1336	0	0	0	0	0	0

*The national average number of workers per household is 1.2. This figure was divided into the number of non-local workers to obtain the number of new households.

**The number of family members per construction household divided by the number of workers per household (1.2) is 1.9. This figure is 2.4 for Dunn County residents. The average of these two figures (2.15) was multiplied by the number of non-local workers to obtain the increase in population.

Table 6.A.3 Projections of In-Migrating Workers, Households, and New Residents in Dunn County, North Dakota

W_i = the capacity of the ith town

d_i = the distance to each i town, and

α = the attractiveness index for commuting

the distribution of new households can be determined. Table 6.A.4 presents the available towns, their populations, their distances from the site, and their projected shares of the new population in the fourth year.

Since most, if not all, of these new residents will be construction workers living in temporary housing units, a significant number of mobile home pads with sewer, water, gas, and electric facilities will be required. In addition, if the impact of the large increases in population is not to cause the social and economic problems associated with declines in local services, significant temporary additions to the public infrastructure of the communities will be required. Table 6.A.5 presents estimates of the expanded services required in Killdeer during the fourth year if the currently prevailing quality of life is to be maintained for all residents. As this table indicates, Killdeer must provide at least three additional temporary classrooms for elementary students and 1.5 temporary classrooms for high school students within one year of the beginning of plant construction. By the fourth year, a total of 9.3 elementary and 4.2 high school classrooms will be needed. However, in the fifth year, these additional classrooms will be obsolete since the temporary construction work force is expected to out-migrate. The temporary nature of the expanded service requirements makes their provision quite difficult. Also, because other neighboring communities are also expected to experience

Table 6.A.4   Spatial Allocation of New Households in Dunn County

Town	Population	Distance	New Households	Percent Increase
Dunn Center	100	2	1160	1160
Killdeer	615	7	582	95
Fayette	100	25	7	7
Manning	100	17	16	16
Halliday	413	15	86	21
Dodge	121	22	12	10
Marshall	100	27	7	7

New Households By Year Service Type	Year 1 (0)	Year 2 (200)	Year 3 (490)	Year 4 (582)	Year 5 (0)
1. Elementary Classrooms	0	3 classrooms	8 classrooms	9.3 classrooms	0
2. High School Classrooms	0	1.5 classrooms	3.5 classrooms	4.2 classrooms	0
3. Recreation Facilities	0	2 acre park .8 acre playgrnd	4 acre park 2 acre playgrnd	4.5 acre park 2 acre playgrnd	0
4. Police Services	0	1 new officer	2 new officers	2 new officers	0
5. Police Facilities	0	200 sq. ft. in station	400 sq. ft. in station	500 sq. ft. in station, 1 new vehicle	0
6. Fire Services	0	none	1 fireman	1 fireman	0
7. Fire Facilities	0	none	200 sq. ft. in station	250 sq. ft. in station	0
8. Solid Waste Services	0	500 man-hrs.	1000 man-hrs.	1100 man-hrs.	0
9. Library Facilities	0	500 sq. ft. in library	1200 sq. ft. in library	1500 sq. ft. in library	0

Source: The requirements for small independent outlying communities were prepared by the Real Estate Research Corporation and are presented in Stenehjem, E.J., and J.E. Metzger, *A Framework for Projecting Employment and Population Changes Accompanying Energy Development, Phase II*, ANL Draft Report, Oct. 1976.

Table 6.A.5 Requirements of Selected Local Services for New Households in Killdeer, North Dakota

"boom-bust" growth, the likelihood of sharing service among communities is low.

The installation of a coal gasification plant in Dunn County is an example of negative socioeconomic impacts. The rapid rate of population growth and decline caused by in- and out-migration of construction worker households raises serious questions as to whether the impacted communities can respond to the increased needs for local services. If the boom-town syndrome exists, this condition may lead to declining living standards, out-migration, and reductions in productivity in both the basic and secondary sectors of the local economy.

In the following appendix, the analyses presented in the Dunn County example will be summarized for a variety of counties and a range of alternative coal technologies. These assessments will demonstrate the sensitivities of socioeconomic impacts to county and technology differences.

APPENDIX B: Empirical Results for the Twenty-Six Cases

YELLOWSTONE COUNTY, MONTANA

TECHNOLOGY: Gasification + 7.5 MM Strip Mine

Multipliers
Simple   3.4
Complex   5.9
Regression 1.7 const. & operation, 1.8 mining.
Use   3.0   Household Factor:

Period	Plant Direct	Mining Direct	Total Basic	Δ In Total Basic	Simple Change Secondary Adjustment Factors 1 (710)	2 (1830)	3 (1600)	4 (500)	5 (−3470)	6 (350)	Δ In Total Sec.	Total Sec.	Δ In Total Jobs	Total Ann. Jobs	Maximum Available Workers Ann. Δ	Cumm. Total	Total Local Employment Ann. Δ	Cumm. Total	Total Non-Local Employment Ann. Δ	Cumm. Total	New Population Ann. Δ	Cumm. Total	New Househld Ann. Δ	Cumm. Totl
1	355		355	355	504						504	504	859	859	1878	1878	859	859	0	0	0	0	0	0
2	1270		1270	915	121	1300					1421	1925	2336	3195	634	2517	1653	2512	683	683	1468	1468	569	569
3	2070		2070	800	57	311	1136				1504	3429	2304	5499	635	3147	635	3147	1669	2352	3588	5056	1390	1959
4	2320		2320	250	28	146	272	355			801	4230	1051	6550	635	3782	635	3782	416	2768	894	5950	347	2306
5	585		585	−1735		73	128	85	−2464		−2178	2052	−3913	2637	640	4422	−1145	2637	−2768	0	−5950	0	−2306	0
6	630	130	760	175			64	40	−590	249	−237	1815	−62	2575	640	5062	−62	2575	0	0	0	0	0	0
7	630	130	760	0				20	−278	60	−198	1617	−198	2377	639	5701	−198	2377	0	0	0	0	0	0
8	630	130	760	0					−139	28	−111	1506	−111	2266	640	6341	−111	2266	0	0	0	0	0	0
9	630	130	760	0						14	+14	1520	+14	2280	639	6980	14	2280	0	0	0	0	0	0
10	630	130	760	0							0	1520	0	2280	639	7619	0	2280	0	0	0	0	0	0

YELLOWSTONE COUNTY, MONTANA

TECHNOLOGY: 800 MW Plant + 2.5MM Strip Mine

Multipliers
Simple   3.4
Complex 5.9
Regression 1.7 const. & operation, 1.8 mining
   Use 3.0 for Household Factor:
      all Basic

Period	Plant Direct	Mining Direct	Total Basic	Δ In Total Basic	Simple Change Secondary / Adjustment Factors 1 (840)	2 (1680)	3 (0)	4 (-1680)	5 (-480)	6	Δ In Total Sec.	Total Sec.	Δ In Total Jobs	Total Ann. Jobs	Maximum Available Workers Ann. Δ	Cumm. Total	Total Local Employment Ann. Δ	Cumm. Total	Total Non-Local Employment Ann. Δ	Cumm. Total	New Population Ann. Δ	Cumm. Total	New Househld Ann. Δ	Cumm. Total
1	420		420	420	595						595	595	1015	1015	1878	1878	1015	1015	0	0	0	0	0	0
2	1200		1260	840	143	1190					1333	1928	2173	3188	634	2512	1497	2512	676	676	1453	1453	563	563
3	1260		1260	0	67	285	0				352	2280	352	3540	635	3147	352	2864	0	676	0	1453	0	563
4	420		420	-840	34	135	0	-1190			-1020	1260	-1860	1680	635	3782	-1184	1680	-676	0	-1453	0	-563	0
5	110	70	180	-240		78	0	-285	-340		-547	713	-787	893	640	4422	-787	893	0	0	0	0	0	0
6	110	70	180	0			0	-135	-82		-217	496	-217	676	640	5062	-217	676	0	0	0	0	0	0
7	110	70	180	0				-78	-38		-116	380	-116	560	639	5701	-116	560	0	0	0	0	0	0
8	110	70	180	0					-20		-20	360	-20	540	640	6341	-20	540	0	0	0	0	0	0
9	110	70	180	0							0	360	0	540	639	6980	0	540	0	0	0	0	0	0
10	110	70	180	0							0	360	0	540	639	7619	0	540	0	0	0	0	0	0

YELLOWSTONE COUNTY, MONTANA

TECHNOLOGY: 6MM Strip Mine

Multipliers
  Simple 3.4
  Complex 5.9
  Regression 1.7 const. 6 operation, 1.8 mining.
  Use 3.0   Household Factor:

Period	Plant Direct	Mining Direct	Total Basic	Δ In Total Basic	200 Simple Change Secondary Adjustment Factors 1	2	3	4	5	6	Δ In Total Sec.	Total Sec.	Δ In Total Jobs	Total Ann. Jobs	Maximum Available Workers Ann. Δ	Cumm. Total	Total Local Employment Ann. Δ	Cumm. Total	Total Non-Local Employment Ann. Δ	Cumm. Total	New Population Ann. Δ	Cumm. Total	New Household Ann. Δ	Cumm. Total
1	100		100	100	142						142	142	242	242	1878	1878	242	242	0	0	0	0	0	0
2			100	0	34						34	176	34	276	634	2517	34	276						
3			100	0	16						16	192	16	292	635	3147	16	292						
4			100	0	8						8	200	8	300	635	3782	8	300						
5			100	0	0						0	200	0	300	640	4422	0	300						
6			100	0	0						0	200	0	300	640	5062	0	300						
7			100	0	0						0	200	0	300	639	5701	0	300						
8			100	0	0						0	200	0	300	640	6341	0	300						
9			100	0	0						0	200	0	300	639	6980	0	300						
10			100	0	0						0	200	0	300	639	7619	0	300						

CUSTER COUNTY, MONTANA

TECHNOLOGY: 800 MW Plant + 2.5 MM Strip Mine

Multipliers
  Simple   3.3
  Complex 5.5
Regression 1.7 (const/op) 1.8 (Mining)
  Use 3.0 for Household Factor:
  all Basic

Period	Plant Direct	Mining Direct	Total Basic	Δ In Total Basic	840 (1)	1680 (2)	0 (3)	-1680 (4)	-480 (5)	(6)	Δ In Total Sec.	Total Sec.	Δ In Total Jobs	Total Ann. Jobs	Max. Avail. Workers Ann. Δ	Max. Avail. Workers Cumm. Total	Total Local Empl. Ann. Δ	Total Local Empl. Cumm. Total	Total Non-Local Empl. Ann. Δ	Total Non-Local Empl. Cumm. Total	New Population Ann. Δ	New Population Cumm. Total	New Household Ann. Δ	New Household Cumm. Total
1	420		420	420	595						595	595	1015	1015	340	340	340	340	675	675	1451	1451	562	562
2	1260		1260	840	143	1190					1333	1928	2173	3188	93	433	93	433	2080	2755	4472	5923	1733	2295
3	1260		1260	0	67	285	0				352	2280	352	3540	98	531	98	531	254	3009	546	6469	211	2506
4	420		420	-840	34	135	0	-1190			-1020	1260	-1860	1680	98	629	0	531	-1860	1149	-4000	2469	-1550	956
5	110	70	180	-240	78		0	-285	-340		-547	713	-787	893	98	727	0	531	-787	362	-1692	777	-655	301
6	110	70	180	0			0	-135	-82		-217	496	-217	676	98	825	0	531	-217	145	-466	311	-180	121
7	110	70	180	0				-78	-38		-116	380	-116	560	93	918	0	531	-116	29	-250	61	-97	24
8	110	70	180	0					-20		-20	360	-20	540	92	1010	0	531	-20	9	-43	18	-16	8
9	110	70	180	0							0	360	0	540	92	1102	0	531	0	9	0	18	0	8
10	110	70	180	0							0	360	0	540	92	1194	0	531	0	9	0	18	0	8

Simple Change Secondary Adjustment Factors (columns 1–6)

ROSEBUD COUNTY

TECHNOLOGY: Gasification + 7.5 MM Strip Mine

Multipliers
Simple 2.2
Complex 3.0
Regression 1.7 constr. & operation, 1.8 mining.
Use 2.2
Household Factor:

Period	Plant Direct	Mining Direct	Total Basic	Δ In Total Basic	Simple Change Secondary / Adjustment Factors 1 (426)	2 (1098)	3 (960)	4 (300)	5 (-2082)	6 (210)	Total Sec.	Δ In Total Sec.	Δ In Total Jobs	Total Ann. Jobs	Max Avail Workers Ann. Δ	Cumm. Total	Total Local Employment Ann. Δ	Cumm. Total	Total Non-Local Employment Ann. Δ	Cumm. Total	New Population Ann. Δ	Cumm. Total	New Household Ann. Δ	Cumm. Total
1	355		355	355	300						300	300	655	655	-240	240	240	240	415	415	870	870	345	345
2	1270		1270	915	70	780					1150	850	1765	2420	70	310	70	310	1695	2110	3560	4430	1410	1760
3	2070		2070	800	35	185	680				2050	900	1700	4120	80	390	80	390	1620	3730	3400	7830	1350	3110
4	2320		2320	250	17	90	160	210			2527	477	730	4850	80	470	80	470	650	4380	1370	9200	540	3650
5	585		585	-1735		45	75	50	-1480		1220	-1310	-3045	1805	80	550	0	470	-3045	1335	-6390	2800	-2540	1110
6	630	130	760	175			35	25	-350	150	1075	-140	35	1840	80	630	35	505	0	1335	0	2800	0	1110
7	630	130	760	0				15	-165	35	945	-130	-130	1710	80	710	0	505	-130	1205	-270	2530	110	1220
8	630	130	760	0					-80	17	960	17	17	1730	75	785	17	522	0	1205	0	2530	0	1220
9	630	130	760	0						8	960	8	0	1730	75	860	0	522	0	1205	0	2530	0	1220
10	630	130	760	0							960	0	0	1730	75	935	0	522	0	1205	0	2530	0	1220

ROSEBUD COUNTY, MONTANA

TECHNOLOGY: 800 MW + Mine

Multipliers
  Simple   2.2
  Complex 3.0
Regression 1.7 (const/op) 1.8 (mining)
  Use   2.2
          Household Factor:

| Period | Plant Direct | Mining Direct | Total Basic | Δ In Total Basic | Simple Change Secondary Adjustment Factors | | | | | | Δ In Total Sec. | Total Sec. | Δ In Total Jobs | Total Ann. Jobs | Maximum Available Workers | | Total Local Employment | | Total Non-Local Employment | | New Population | | New Household | |
					505 (1)	1010 (2)	0 (3)	-1010 (4)	-290 (5)	6					Ann. Δ	Cumm. Total	Ann. Δ	Cumm. Total	Ann. Δ	Cumm. Total	Ann. Δ	Cumm. Total	Ann. Δ	Cumm. Total
1	420		420	420	360						360	360	780	780	240	240	240	240	540	540	1130	1130	450	450
2	1260		1260	840	85	720					805	1165	1645	2425	70	310	70	310	1575	2115	3310	4440	1310	1760
3	1260		1260	0	40	170	0				210	1375	210	2635	80	390	80	390	130	2245	270	4710	110	1870
4	420		420	-840	20	80	0	-720			-620	755	-1460	1175	80	470	0	390	-1460	785	-3060	1650	-1220	650
5	110	70	180	-240		40	0	-170	-205		-335	420	-575	600	80	550	0.	390	-575	210	-1200	450	-480	170
6	110	70	180	0			0	-80	-50		-130	290	-130	470	80	630	0	390	-130	80	-270	180	-105	65
7	110	70	180	0				-40	-25		-65	225	-65	405	80	710	0	390	-65	15	-140	40	-50	15
8	110	70	180	0					-10		-10	215	-10	395	75	785	0	390	-10	5	-20	20	-10	5
9	110	70	180	0							0	215	0	395	75	860	0	390	0	5	0	20	0	5
10	110	70	180	0							0	215	0	395	75	935	0	390	0	5	0	20	0	5

ROSEBUD COUNTY, MONTANA

TECHNOLOGY: 6MM Strip Mine

Multipliers
  Simple 2.2
  Complex 3.0
Regression 1.7 const. & operation, 1.8 mining.
  Use 2.2 Household Factor:

Period	Plant Direct	Mining Direct	Total Basic	Δ In Total Basic	120 Simple Change Secondary (Adjustment Factors) 1	2	3	4	5	6	Δ In Total Sec.	Total Sec.	Δ In Total Jobs	Total Ann. Jobs	Maximum Available Workers Ann. Δ	Cumm. Total	Total Local Employment Ann. Δ	Cumm. Total	Total Non-Local Employment Ann.	Cumm. Total	New Population Ann.	Cumm. Total	New Househld Ann	Cumm Totl
1	100	100	100	100	85						85	85	185	185	240	240	185	185	0	0	0	0	0	0
2	100	100		0	20						20	105	20	205	70	310	20	205						
3	100	100		0	10						10	115	10	215	80	390	10	215						
4	100	100		0	5						5	120	5	220	80	470	5	220						
5	100	100		0							0	120	0	220	80	550	0	220						
6	100	100		0							0	120	0	220	80	630	0	220						
7	100	100		0							0	120	0	220	80	710	0	220						
8	100	100		0							0	120	0	220	75	785	0	220						
9	100	100		0							0	120	0	220	75	785	0	220						
10															75	935	0	220						

PEORIA COUNTY, ILLINOIS

TECHNOLOGY: Coal-Fired P Plants (1600 MWe) with Mines (2 2MM, Strip)

Multipliers
Simple   2.4
Complex   2.4
Regression 1.5 & 1.6 Manuf. & Mining
Use   2.4 Household Factor:

Simple Change Secondary Adjustment Factors column totals: 1 = 588, 2 = 1176, 3 = 588, 4 = 0, 5 = -448, 6 = -1316

Period	Plant Direct	Mining Direct	Total Basic	Δ In Total Basic	Adj. Factor 1	2	3	4	5	6	Δ In Total Sec.	Total Sec.	Δ In Total Jobs	Total Ann. Jobs	Max Avail Workers Ann. Δ	Max Avail Workers Cumm. Total	Total Local Employment Ann. Δ	Total Local Employment Cumm. Total	Total Non-Local Employment Ann. Δ	Total Non-Local Employment Cumm. Total	New Population Ann. Δ	New Population Cumm. Total	New Household Ann. Δ	New Household Cumm. Total
1	420		420	420	417						417	417	837	837	3862	3862	837	837						
2	1260		1260	840	100	835					935	1352	1775	2612	1336	5198	1775	2612						
3	1680		1680	420	47	200	417				664	2016	1084	3696	1404	6602	1084	3696						
4	1680		1680	0	24	94	100	0			218	2234	218	3914	1403	8005	218	3914						
5	1260	100	1360	-320		47	47	0	-318		-224	2010	-544	3370	1402	9407	-544	3370						
6	420	100	520	-840		0	24	0	-76	-835	-887	1123	-1727	1643	1402	10809	-1727	1643						
7	220	200	420	-100			0	0	-36	-299	-335	788	-435	688	1363	12172	-435	688						
8	220	200	420	0			0	0	-18	-118	-136	652	-136	652	1329	13501	-136	652						
9	220	200	420	0				0	0	-58	-58	594	-116	594	1328	14829	-116	594						
10	220	200	420	0						-6	-6	588	-12	588	1327	16156	-12	588						
11	220	200	420	0						0	0	588	0	588			0	588						

PEORIA COUNTY, ILLINOIS

TECHNOLOGY: Coal-Fired P Plant (800 MWe) with Mine (2MM, Strip)

Multipliers
Simple 2.4
Complex
Regression 1.5 & 1.6 Manuf. & Mining
Use 2.4 Household Factor:

Period	Plant Direct	Mining Direct	Total Basic	Δ In Total Basic	Simple Change Secondary — Adjustment Factors 1 (588)	2 (1176)	3 (0)	4 (−1176)	5 (−294)	6	Δ In Total Sec.	Total Sec.	Δ In Total Jobs	Total Ann. Jobs	Maximum Available Workers Ann.Δ	Cumm. Total	Total Local Employment Ann.Δ	Cumm. Total	Total Non-Local Employment Ann.Δ	Cumm. Total	New Population Ann.Δ	Cumm. Total	New Household Ann.Δ	Cumm. Total
1	420		420	420	417						417	417	837	837	3862	3862	837	837		837				
2	1260		1260	840	100	835					935	1352	1775	2612	1336	5198	1775	2612		2612				
3	1260		1260	0	47	200	0				247	1599	247	2859	1404	6602	247	2859		2859				
4	420		420	−840	24	94	0	−835			−717	882	−1557	1302	1403	8005	−1557	1302		1302				
5	110	100	210	−210	0	47	0	−200	−201		−354	528	−564	738	1402	9407	−564	738		738				
6	110	100	210	0	0	0	0	−94	−50		−144	384	−144	594	1402	10809	−144	594		594				
7	110	100	210	0		0	0	−47	−24		−71	313	−71	523	1363	12172	−71	523		523				
8	110	100	210	0			0	0	−12		−12	301	−12	511	1329	13501	−12	511		511				
9	110	100	210	0				0			0	301	0	511	1328	14829	0	511		511				
10	110	100	210	0				0			0	301	0	511	1327	16156	0	511		511				
11																								
12																								
15	110	100	210	0								301	511	511										
20	110	100	210	0								301	511	511										
25	110	100	210	0								301	511	511										
30	110	100	210	0								301	511	511										

PEORIA COUNTY, ILLINOIS

TECHNOLOGY: Coal-Fired P Plant (800 MWe) without Mine

Multipliers
  Simple   2.4
  Complex
  Regression 1.4 & 1.6 Manuf. & Mining
  Use   2.4   Household Factor:

Simple Change Secondary — Adjustment Factors header coefficients: 588 1176 0 -1176 -434

Period	Plant Direct	Mining Direct	Total Basic	Δ In Total Basic	1	2	3	4	5	6	Δ In Total Sec.	Total Sec.	Δ In Total Jobs	Total Ann. Jobs	Max. Avail. Workers Ann. Δ	Max. Avail. Workers Cumm. Total	Total Local Employment Ann. Δ	Total Local Employment Cumm. Total	Total Non-Local Employment Ann. Δ	Total Non-Local Employment Cumm. Total	New Population Ann. Δ	New Population Cumm. Total	New Household Ann. Δ	New Household Cumm. Total
1			420	420	417						417	417	837	837	3862	3862	837	837						
2			1260	840	100	835					935	1352	1775	2612	1336	5198	1775	2612						
3			1260	0	47	200	0				247	1599	247	2859	1404	6602	247	2859						
4			420	-840	24	94	0	-835			-717	882	-1557	1302	1403	8005	-1557	1302						
5			110	-310	0	47	0	-200	-308		-461	421	-771	531	1402	9407	-771	531						
6			110	0	0	0	0	-94	-74		-168	253	-168	363	1402	10,809	-168	363						
7			110	0		0	0	-47	-35		-82	171	-82	281	1363	12,172	-82	281						
8			110	0			0	0	-17		-17	154	-17	264	1329	13,501	-17	264						
9			110	0				0	0		0	154	0	264	1328	14,829	0	264						
10			110												1327	16,156								

PERRY COUNTY, ILLINOIS

TECHNOLOGY: Gasification with Strip

Multipliers 1.8
  Simple 1.8
  Comples
  Regression 1.5 & 1.6 Manuf. & Mining
  Use 1.8 Household Factor:

Period	Plant Direct	Mining Direct	Total Basic	Δ In Total Basic	284 (1)	732 (2)	640 (3)	200 (4)	-1788 (5)	180 (6)	Δ In Total Sec.	Total Sec.	Δ In Total Jobs	Total Ann. Jobs	Max. Avail. Workers Ann. Δ	Max. Avail. Workers Cumm. Total	Total Local Emp. Ann. Δ	Total Local Emp. Cumm. Total	Non-Local Emp. Ann. Δ	Non-Local Emp. Cumm. Total	New Pop. Ann. Δ	New Pop. Cumm. Total	New HH Ann. Δ	New HH Cumm. Total
1	355		355	355	202						202	202	557	557	4008	4008	557	557				0		0
2	1270		1270	915	48	520					568	770	1483	2040	473	4481	1483	2040						
3	2070		2070	800	23	124	454				601	1371	1401	3441	385	4866	1401	3441						
4	2320		2320	250	11	59	109	142			321	1692	571	4012	385	5251	571	4012						
5	585		585	-1735	0	29	51	34	-985		-871	821	-2606	1406	385	5636	-2606	1406				0		0
6	630	180	810	225		0	26	16	-236	128	-64	757	161	1567	321	6007	161	1567						
7	630	180	810	0			0	8	-111	31	-71	686	-71	1496	364	6371	-71	1496						
8	630	180	810	0				0	-56	14	-41	645	-41	1455	364	6735	-41	1455				0		
9	630	180	810	0					0	7	7	652	7	1462	364	7099	7	1462						
10	630	180	810	0						0	0	652	0	1462	363	7462	0	1462						

PERRY COUNTY, ILLINOIS

TECHNOLOGY: 2 800MW Plants constructed simultaneously with 2, 2M Tons/Yr strip mines

Multipliers
Simple 1.8
Complex 2.3
Regression 1.5 construction
Use 1.8 Household Factor:

Period	Plant Direct	Mining Direct	Total Basic	Δ In Total Basic	Simple Change Secondary Adjustment Factors 1	2	3	4	5	6	Δ In Total Sec.	Total Sec.	Δ In Total Jobs	Total Ann. Jobs	Max. Avail. Workers Ann. Δ	Cumm. Total	Total Local Empl. Ann. Δ	Cumm. Total	Total Non-Local Empl. Ann. Δ	Cumm. Total	New Population Ann. Δ	Cumm. Total	New Household Ann. Δ	Cumm. Total
					1312	1376	-1376	-976																
1	1640		1640	1640	931						931	931	2571	2571	4008	4008	2571	2571	0	0	0	0	0	0
2	3360		3360	1720	223	976					1199	2130	2919	5490	473	4481	1910	4481	1009	1009	2220	2220	840	840
3	1640		1640	-1720	105	234	-976				-637	1493	-2357	3133	385	4866	-1348	3133	-1009	0	-2220	0	-840	0
4	220	200	420	-1220	52	110	-234	-693			-765	728	-1985	1148	385	5251	-1985	1148	0	0	0	0	0	0
5	220	200	420	0		55	-110	-166			-221	507	-221	927	385	5636	-221	927	0	0	0	0	0	0
6	220	200	420	0			55	-55	-78		-133	374	-133	794	371	6007	-133	794	0	0	0	0	0	0
7	220	200	420	0					-39		-39	335	-39	755	364	6371	-39	755	0	0	0	0	0	0
8	220	200	420	0							0	335	0	755	364	6735	0	755	0	0	0	0	0	0
9	220	200	420	0							0	335	0	755	364	7099	0	755	0	0	0	0	0	0
10	220	200	420	0							0	335	0	755	364	7462	0	755	0	0	0	0	0	0

PERRY COUNTY, ILLINOIS

TECHNOLOGY: Coal-Fired P Plants (1600 MWe) with Mines (2 2MM, Strip)

Multipliers
Simple 1.8
Complex
Regression 1.5 & 1.6 Manuf. & Mining
Use   Household Factor:

Period	Plant Direct	Mining Direct	Total Basic	Δ In Total Basic	336 (1)	672 (2)	336 (3)	0 (4)	-256 (5)	-752 (6)	Δ In Total Sec.	Total Sec.	Δ In Total Jobs	Total Ann. Jobs	Max. Avail. Workers Ann. Δ	Max. Avail. Workers Cumm. Total	Total Local Emp. Ann. Δ	Total Local Emp. Cumm. Total	Non-Local Emp. Ann. Δ	Non-Local Emp. Cumm. Total	New Pop. Ann. Δ	New Pop. Cumm. Total	New HH Ann. Δ	New HH Cumm. Total
1	420		420	420	239						239	239	659	659	4008	4008	659	659			0		0	
2	1260		1260	840	57	447					504	743	1344	2003	473	4481	1344	2003						
3	1680		1680	420	27	114	239				380	1123	800	2803	385	4866	800	2803						
4	1680		1680	0	13	54	57	0			124	1247	124	2927	385	5251	124	2927						
5	1260	100	1360	-320	0	27	27	0	-182		-128	1119	-448	2479	385	5636	-448	2479			0		0	
6	420	100	520	-840	0	0	13	0	-44	-447	-478	641	-1318	1161	371	6007	-1318	1161						
7	220	200	420	-100			0	0	-20	-171	-191	450	-291	870	364	6371	-291	870						
8	220	200	420	0				0	-10	-68	-78	372	-78	792	364	6735	-78	792			0		0	
9	220	200	420	0					0	-33	-33	339	-33	759	364	7099	-33	759						
10	220	200	420	0						-3	-3	336	-3	756	363	7462	-3	756						

(Simple Change — Adjustment Factors)
(Secondary)

PERRY COUNTY, ILLINOIS

TECHNOLOGY: Coal-Fired P Plant (800 MWe) with Mine (2MM, Strip)

Multipliers
Simple 1.8
Complex
Regression 1.5 & 1.6 Manuf. & Mining
Use 1.8   Household Factor:

Period	Plant Direct	Mining Direct	Total Basic	Δ In Total Basic	336 (1)	672 (2)	0 (3)	-672 (4)	-168 (5)	(6)	Δ In Total Sec.	Total Sec.	Δ In Total Jobs	Total Ann. Jobs	Max. Avail. Workers Ann. Δ	Cumm. Total	Total Local Empl. Ann. Δ	Cumm. Total	Total Non-Local Empl. Ann. Δ	Cumm. Total	New Population Ann. Δ	Cumm. Total	New Household Ann. Δ	Cumm. Total
									Simple Change Secondary Adjustment Factors															
1	420		420	420	289						239	239	659	659	4008	4008	659	659			0	0	0	0
2	1260		1260	840	57	477					534	773	1374	2033	473	4481	1374	2033						
3	1260		1260	0	27	114	0				141	914	141	2174	385	4866	141	2174						
4	420		420	-840	13	54	0	-477			-464	450	-1304	870	385	5251	-1304	870						
5	110	100	210	-210	0	27	0	-114	-119		-206	244	-416	454	385	5636	-416	454			0	0	0	0
6	110	100	210	0	0	0	0	-54	-29		-83	161	-83	371	371	6007	-83	371						
7	110	100	210	0	0	0	0	-27	-13		-40	121	-40	331	364	6371	-40	331						
8	110	100	210	0			0	0	-7		-7	114	-7	324	364	6735	-7	324			0	0	0	0
9	110	100	210	0							0	114	0	324	364	7099	0	324						
10	110	100	210	0							0	114	0	324	363	7462	0	324						

PERRY COUNTY, ILLINOIS

TECHNOLOGY: Coal Mine 2-3MM Strip

Multipliers
  Simple 1.8
  Complex
Regression 1.5 & 1.6 Manuf. & Mining
  Use   Household Factor:

Period	Plant Direct	Mining Direct	Total Basic	Δ In Total Basic	Simple Change Secondary Adjustment Factors 1	2	3	4	5	6	Δ In Total Sec.	Total Sec.	Δ In Total Jobs	Total Ann. Jobs	Maximum Available Workers Ann. Δ	Cumm. Total	Total Local Employment Ann. Δ	Cumm. Total	Total Non-Local Employment Ann. Δ	Cumm. Total	New Population Ann. Δ	Cumm. Total	New Household Ann. Δ	Cumm. Total
1	240		240	240	136						136	136	376	376	4008	4008	376	376	0	0				
2		240	240	0	33		0				33	169	33	409	473	4481	33	409	0	0				
3				0	15						15	184	15	424	385	4866	15	424	0	0				
4					8						8	192	8	432	385	5251	8	432	0	0				
5					0						0	192	0	432	385	5636	0	432	0	0				

HAMILTON COUNTY, ILLINOIS

TECHNOLOGY: Gasification with Deep Mine

Multipliers
  Simple 2.0
  Complex
  Regression 1.5 & 1.6 Manuf. & Mining
  Use 2.0, Household Factor: 2.0 & 1.9

Simple Change Secondary Adjustment Factors column inputs (Δ In Total Basic): 355, 915, 800, 250, -1735, 1305

Period	Plant Direct	Mining Direct	Total Basic	Δ In Total Basic	Simple Change Secondary Adjustment Factors 1	2	3	4	5	6	Δ In Total Sec.	Total Sec.	Δ In Total Jobs	Total Ann. Jobs	Maximum Available Workers Ann. Δ	Cumm. Total	Total Local Employment Ann. Δ	Cumm. Total	Total Non-Local Employment Ann. Δ	Cumm. Total	New Population Ann. Δ	Cumm. Total	New Household Ann. Δ	Cumm. Total
1	355		355	355	252						252	252	607	607	2284	2284	607	607	0	0	0	0	0	0
2	1270		1270	915	60	650					710	962	1625	2232	261	2545	1625	2232	0	0	0	0	0	0
3	2070		2070	800	28	156	568				752	1714	1552	3784	292	2837	605	2837	947	947	1894	1894	789	789
4	2320		2320	250	14	73	136	178			401	2115	651	4435	292	3129	292	3129	359	1306	718	2612	299	1088
5	585		585	-1735	0	37	64	43	-1232		-1088	1027	-2823	1612	292	3421	-1517	1612	-1306	0	-2612	0	-1088	0
6	630	1260	1890	1305		0	32	20	-295	927	684	1711	1989	3601	291	3712	1803	3420	181	181	362	362	151	151
7	630	1260	1890	0			0	10	-139	222	83	1794	83	3684	281	3993	83	3503	0	181	0	362	0	151
8	630	1260	1890	0				0	-69	104	35	1829	35	3719	259	4252	35	3538	0	181	0	362	0	151
9	630	1260	1890	0					0	52	52	1889	52	3771	258	4510	52	3590	0	181	0	362	0	151
10	630	1260	1890	0						0	0	0	0	3771	258	4768	0	3590	0	181	0	362	0	151
											0		0	3771										
											0		0	3771										

HAMILTON COUNTY, ILLINOIS

TECHNOLOGY: Coal Fired Plant, 800 MWe, with Mine (2MM, Deep)

Multipliers
  Simple 2.0
  Complex
  Regression 1.5 & 1.6 Manuf. & Mining
  Use 2.0   Household Factor:

Period	Plant Direct	Mining Direct	Total Basic	Δ In Total Basic	Simple Change Secondary Adjustment Factors 1	2	3	4	5	6	Δ In Total Sec.	Total Sec.	Δ In Total Jobs	Total Ann. Jobs	Max. Avail. Workers Ann. Δ	Max. Avail. Workers Cumm. Total	Total Local Employment Ann. Δ	Total Local Employment Cumm. Total	Total Non-Local Employment Ann. Δ	Total Non-Local Employment Cumm. Total	New Population Ann. Δ	New Population Cumm. Total	New Household Ann. Δ	New Household Cumm. Total
1	420		420	420	298						298	298	718	718	2284	2284	718	718	0	0				
2	1260		1260	840	71	596					667	965	1507	2225	261	2545	1507	2225	0	0				
3	1260		1260	0	34	143	0				177	1142	177	2402	292	2837	177	2402						
4	420		420	-840	17	67	0	-598			-512	630	-1352	1050	292	3129	-1352	1050						
5	110	425	535	115	0	34	0	-143	82		-27	603	88	1138	292	3421	88	1138						
6	110	425	535	0	0	0	0	-67	20		-47	556	-47	1091	291	3712	-47	1091						
7	110	425	535	0	0	0	0	-34	9		-25	531	-25	1066	281	3993	-25	1066						
8	110	425	535	0				0	0	4	4	535	4	1070	259	4252	4	1070						
9	110	425	535	0						0	0	535	0	1070	258	4510	0	1070						
10	110	425	535	0						0	0	535	0	1070	258	4768	0	1070						

HAMILTON COUNTY, ILLINOIS

TECHNOLOGY: Coal Mines (3-2 MM, Deep)

Multipliers
  Simple 2.0
  Complex
  Regression 1.5 & 1.6 Manuf. & Mining
  Use 2.0 Household Factor:

Period	Plant Direct	Mining Direct	Total Basic	Δ In Total Basic	Simple Change Secondary Adjustment Factors 1	2	3	4	5	6	Δ In Total Sec.	Total Sec.	Δ In Total Jobs	Total Ann. Jobs	Maximum Available Workers Ann.	Cumm. Total	Total Local Employment Ann.	Cumm. Total	Total Non-Local Employment Ann.	Cumm. Total	New Population Ann.	Cumm. Total	New Household Ann.	Cumm. Total
1		1275	1275	1275	905						905	905	2180	2180	2284	2284	2180	2180						
2		1275	1275	0	217	0					217	1122	217	2397	261	2545	217	2397						
3		1275	1275	0	102	0					102	1204	102	2499	292	2837	102	2499						
4				51							51	1255	51	2550	292	3129	51	2550						
5		1275		0	0						0	1255	0	2550	292	3421	0	2550						

DUNN COUNTY, NORTH DAKOTA

TECHNOLOGY: Gasification + 9MM Strip Mine

Multipliers
Simple 1.6
Complex 1.7
Regression 1.5 Mining, 1.6 Const. + Oper.
Use 1.6
Household Factor: 2.4 + 1.9

Period	Plant Direct	Mining Direct	Total Basic	Δ In Total Basic	210 (1)	550 (2)	480 (3)	150 (4)	-1040 (5)	150 (6)	Δ In Total Sec.	Total Sec.	Δ In Total Jobs	Total Ann. Jobs	Max. Avail. Workers Ann. Δ	Max. Avail. Workers Cumm. Total	Total Local Empl. Ann. Δ	Total Local Empl. Cumm. Total	Total Non-Local Empl. Ann. Δ	Total Non-Local Empl. Cumm. Total	New Population Ann. Δ	New Population Cumm. Total	New Household Ann. Δ	New Household Cumm. Total
								Simple Change Secondary Adjustment Factors																
1	355		355	355	150						150	150	505	505	950	950	505	505	0	0	0	0	0	0
2	1270		1270	915	35	390					425	575	1340	1895	120	1070	505	1070	775	775	1705	1705	645	645
3	2070		2070	800	18	90	340				448	1023	1248	3090	135	1250	135	1205	1115	1890	2450	4155	930	1575
4	2320		2320	250	10	45	80	105			240	1263	490	3580	135	1340	135	1340	355	2245	780	4935	295	1870
5	585		585	-1735		23	40	25	-740		-650	613	-2385	1200	135	1475	-140	1200	-2245	0	-4935	0	-1870	0
6	630	200	830	245			20	12	-175	105	-38	575	207	1407	135	1610	135	1335	72	72	160	160	60	60
7	630	200	830	0				6	-80	25	-49	526	-49	1358	125	1860	0	1335	-49	23	-110	40	-40	20
8	630	200	830	0					-40	12	-28	498	-28	1330	115	1975	-5	1330	-23	0	-50	0	-20	0
9	630	200	830	0						6	6	492	6	1336	115	2090	6	1336	0	0	0	0	0	0
10	630	200	830	0							0	492	0	1336	115	2250	0	1336	0	0	0	0	0	0

DUNN COUNTY

TECHNOLOGY: 1600 MW + Mine

Multipliers
  Simple 1.6
  Complex 1.7
  Regression 1.5 mining, 1.6 constr. & operation
  Use 1.6 Household Factor:

Period	Plant Direct	Mining Direct	Total Basic	Δ In Total Basic	*250 (1)	*500 (2)	*250 (3)	*0 (4)	*-250 (5)	*-560 (6)	Δ In Total Sec.	Total Sec.	Δ In Total Jobs	Total Ann. Jobs	Max. Avail. Workers Ann. Δ	Max. Avail. Workers Cumm. Total	Total Local Emp. Ann. Δ	Total Local Emp. Cumm. Total	Total Non-Local Emp. Ann. Δ	Total Non-Local Emp. Cumm. Total	New Population Ann. Δ	New Population Cumm. Total	New Household Ann. Δ	New Household Cumm. Total
1	420		420	420	170						170	170	590	590	950	950	590	590	0	0	0	0	0	0
2	1260		1260	840	40	350					390	560	1230	1820	120	1070	480	1070	750	750	1575	1575	625	625
3	1680		1680	420	20	80	170				270	830	690	2510	135	1205	135	1205	555	1305	1165	2740	463	1088
4	1680		1680	0	10	40	40	0			90	920	90	2600	135	1340	90	1295	0	1305	0	2740	0	1088
5	1260		1260	-420		20	20	0	-170		-130	790	-550	2050	135	1475	0	1295	-550	755	-1155	1585	-458	630
6	420		420	-840			10	0	-40	-350	-380	410	-1220	830	135	1610	-465	830	-755	0	-1585	0	-630	0
7	220	100	320	-100				0	-20	-120	-140	270	-240	590	250	1860	-240	590	0	0	0	0	0	0
8	220	100	320	0					-10	-50	-60	210	-60	530	115	1975	-60	530	0	0	0	0	0	0
9	220	100	320	0						-25	-25	185	-25	505	115	2090	-25	505	0	0	0	0	0	0
10	220	100	320	0						-5	-5	180	-5	500	115	2205	-5	500	0	0	0	0	0	0
														500			0	500	0	0	0	0	0	0

Simple Change Secondary Adjustment Factors (columns 1–6 header values: 250, 500, 250, 0, -250, -560)

DUNN COUNTY

TECHNOLOGY: 800 MW + Mine

Multipliers
  Simple 1.6
  Complex 1.7
  Regression 1.5 mining, 1.6 constr. & operation
  Use 1.6 Household Factor:

Period	Plant Direct	Mining Direct	Total Basic	Δ In Total Basic	Simple Change Secondary Adjustment Factors 250 / 1	500 / 2	0 / 3	-500 / 4	-140 / 5	6	Δ In Total Sec.	Total Sec.	Δ In Total Jobs	Total Ann. Jobs	Max Avail Workers Ann. Δ	Max Avail Workers Cumm. Total	Total Local Employment Ann. Δ	Total Local Employment Cumm. Total	Total Non-Local Employment Ann. Δ	Total Non-Local Employment Cumm. Total	New Population Ann. Δ	New Population Cumm. Total	New Household Ann. Δ	New Household Cumm. Total
1	420		420	420	175						175	175	595	595	950	950	595	595	0	0	0	0	0	0
2	1260		1260	840	45	350					395	570	1235	1830	120	1070	475	1070	760	760	1600	1600	630	630
3	1260		1260	0	20	80	0				100	670	100	1930	135	1205	100	1170	0	0	0	0	0	0
4	420		420	-840	10	40	0	-350			-300	310	-1140	790	135	1340	-380	790	-760	0	-1000	0	-630	0
5	110	75	185	-235		20	0	-80	-100		-100	210	-395	395	135	1475	-395	395	0	0	0	0	0	0
6	110	75	185	0			0	-40	-20		-60	150	-60	335	135	1610	-60	335	0	0	0	0	0	0
7	110	75	185	0				-20	-10		-30	120	-30	305	125	1860	-30	305	0	0	0	0	0	0
8	110	75	195	0					-5		-5	115	-5	300	115	1975	-5	300	0	0	0	0	0	0
9			185	0								115	0	300	115	2090	0	300	0	0	0	0	0	0
10			185									115		300	115	2205	0	300	0	0	0	0	0	0

DUNN COUNTY

TECHNOLOGY: 6 MM Strip Mine
Multipliers
  Simple 1.6
  Complex 1.7
  Regression 1.5 mining, 1.6 constr. & Operation
  Use 1.6
  Household Factor:

Period	Plant Direct	Mining Direct	Total Basic	Δ In Total Basic	61 [1] Simple Change Secondary Adjustment Factors						Δ In Total Sec.	Total Sec.	Δ In Total Jobs	Total Ann. Jobs	Maximum Available Workers Ann. Δ	Cumm. Total	Total Local Employment Ann. Δ	Cumm. Total	Total Non-Local Employment Ann. Δ	Cumm. Total	New Population Ann. Δ	Cumm. Total	New Household Ann. Δ	Cumm. Total
					1	2	3	4	5	6														
1	100		100	100	40						40	40	140	140	950	950	140	140	0	0	0	0	0	0
2	100		100	0	15						15	55	15	155	120	1070	15	155						
3	100		100	0	5						5	60	5	160	135	1205	5	160						
4	100		100	0	1						1	60	1	160	135	1340	1	161						
5	100		100	0	0						0	60	0	160	135	1475	0	161						
6	100		100	0	0						0	60	0	160	135	1610	0	161						
7	100		100	0	0						0	60	0	160	125	1860	0	161						
8	100		100	0	0						0	60	0	160	115	1975	0	161						

MONONGALIA COUNTY, W. VA.

TECHNOLOGY: Coal Gasification with Mines (2 2MM + 1MM, Deep)

Multipliers
Simple   3.1
Complex
Regression
Use   3.1

Household Factor:

Period	Plant Direct	Mining Direct	Total Basic	Δ In Total Basic	746 (1)	1922 (2)	1680 (3)	525 (4)	-3644 (5)	2405 (6)	Δ In Total Sec.	Total Sec.	Δ In Total Jobs	Total Ann. Jobs	Max. Avail. Workers Ann. Δ	Max. Avail. Workers Cumm. Total	Total Local Employment Ann. Δ	Total Local Employment Cumm. Total	Total Non-Local Employment Ann. Δ	Total Non-Local Employment Cumm. Total	New Population Ann. Δ	New Population Cumm. Total	New Household Ann. Δ	New Household Cumm. Total
1	355		355	355	529						529	529	884	884	6353	6353	884	884						
2	1270		1270	915	127	1364					1491	2020	2406	3290	703	7056	2406	3290						
3	2070		2070	800	60	327	1193				1580	3600	2380	5670	344	7400	2380	5670						
4	2320		2320	250	30	154	286	373			843	4443	1093	6763	343	7743	1093	6763						
5	585		585	-1735	0	77	134	89	-2587		-2287	2183	-3977	2786	343	8086	-3977	2786						
6	630	1100	1730	1145	0	0	67	42	-619	1707	1197	3380	2324	5110	343	8429	2324	5110						
7	630	1100	1730	0	0	0	0	21	-291	409	139	3519	139	5249	337	8766	139	5249						
8	630	1100						0	-146	192	+48	3565	46	5295	301	9067	46	5295						
9	630	1100								96	96	3661	96	5391	301	9368	96	5391						
10										0	0	3661	0	5391	300	9668	0	5391						

MONONGALIA COUNTY, W. VA.

TECHNOLOGY: Coal-Fired P Plant (800 MWe) with Mine (1.9MM, Deep)

Multipliers
Simple 3.1
Complex
Regression
Use 3.1

Household Factor:

Period	Plant Direct	Minint Direct	Total Basic	Δ In Total Basic	Simple Change Secondary Adjustment Factors						Δ In Total Sec.	Total Sec.	Δ In Total Jobs	Total Ann. Jobs	Maximum Available Workers Ann. Δ	Cumm. Total	Total Local Employment Ann. Δ	Cumm. Total	Total Non-Local Employment Ann. Δ	Cumm. Total	New Population Ann. Δ	Cumm. Total	New Household Ann. Δ	Cumm. Total
					882	1764	0	-1764	242															
					1	2	3	4	5	6														
1	420		420	420	626						626	626	1046	1046	6353	6353	1046	1046						
2	1260		1260	840	150	1252					1402	2028	2242	3288	703	7056	2242	3288						
3	1260		1260	0	71	300	0				371	2399	371	3659	344	7400	371	3659						
4	420		420	-840	35	142	0	-1252			-1110	1289	-1950	1709	343	7743	-1950	1709						
5	110	425	535	115	0	70	0	-300	171		-59	1230	56	1765	343	8086	56	1765						
6	110	425	535	0	0	0	0	-142	41		-101	1129	-101	1664	343	8429	-101	1664						
7	110	425	535	0	0	0	0	-70	19		-51	1078	-51	1613	337	8766	-51	1613						
8	110	425						0	10		10	1088	10	1623	301	9067	10	1623						
9	110	425	535	0					0		0	1088	0	1623	301	9368	0	1623						
10	110	425	535	0					0		0	1088	0	1623	300	9668	0	1623						

144

MONONGALIA COUNTY, W. VA.

TECHNOLOGY: Deep Coal Mines (6 1MM)
Multipliers 3.1 Ratiom 1.1 Manuf. & Mining
  Simple 3.1
  Complex
  Regression
  Use 3.1   Household Factor:

| | | | | 3150 | 0 | | | | | | Maximum Available Workers | | Total Local Employment | | Total Non-Local Employment | | New Population | | New Household | |
Period	Plant Direct	Mining Direct	Total Basic	Δ In Total Basic	1	Simple Change Secondary 2 3 4 5 6 Adjustment Factors	Δ In Total Sec.	Total Sec.	Δ In Total Jobs	Total Ann. Jobs	Ann. Δ	Cumm. Total	Ann. Δ	Cumm. Total	Ann. Δ	Cumm. Total	Ann. Δ	Cumm. Total	Ann. Δ	Cumm. Total
1		1500	1500	1500	2236		2236	2236	3736	3736	6353	6353	3736	3736						
2		1500	0	0	536		536	2772	536	4272	703	7056	536	4272						
3		1500	0	0	252		252	3024	252	4524	344	7400	252	4524						
4		1500	0	0	126		126	3150	126	4650	343	7743	126	4650						

WYOMING, W. VA.

TECHNOLOGY: Coal-Fired P Plant (800 MWe) with Mine (1.9MM, Deep)
Multipliers
Simple 1.6
Complex
Regression 1.1 for Manuf. & Mining.
Use Household Factor:

Period	Plant Direct	Mining Direct	Total Basic	Δ In Total Basic	Simple Change Secondary Adjustment Factors 1 (252)	2 (504)	3 (0)	4 (-504)	5 (69)	6	Δ In Total Sec.	Total Sec.	Δ In Total Jobs	Total Ann. Jobs	Maximum Available Workers* Ann. Δ	Maximum Available Workers* Cumm Total	Total Local Employment Ann. Δ	Total Local Employment Cumm. Total	Total Non-Local Employment Ann. Δ	Total Non-Local Employment Cumm. Total	New Population Ann. Δ	New Population Cumm. Total	New Household Ann. Δ	New Household Cumm. Total
1	420		420	420	179						179	179	599	599	3722	3722	599	599						
2	1260		1260	840	43	358					401	580	1241	1840	261	3983	1241	1840						
3	1260		1260	0	20	86	0				106	686	106	1946	290	4273	106	946						
4	420		420	-840	10	40	0	-358			-308	378	-1148	798	290	4563	-1148	798						
5	110	425	535	115	0	20	0	-86	49		-17	361	98	896	290	4853	98	896						
6	110	425	535	0		0	0	-40	12		-28	333	-28	868	290	5143	-28	868						
7	110	425	535				0	-20	6		-14	319	-14	854	280	5423	-14	854						
8	110	425	535					0	3		3	322	3	857	224	5647	3	857						
9	110	425	535						0		0	0	0	857	224	5871	0	857						
10	110	425	535						0		0	0	0	857	224	6095	0	857						

*Labor Force Figures are for Wyoming County only.

NOTES

1. The authors would like to acknowledge the assistance of James Metzger in the preparation of many of the data presented in this report. Special thanks are also due to Thomas Baldwin and Donald Johnson, who assisted the authors in the selection and characterization of sites and technologies. We would also like to thank June Johansen for her untiring efforts to compile, type, edit, and piece together this document.

2. Summary tables of expected annual population and employment changes associated with each type of coal development in each of nine counties are presented in Appendix B. There is one table for each of the 26 separate cases considered.

3. The gravity model predicts that of the communities within easy commuting distance of plant site in Southwestern Yellowstone County, only Billings and Laurel will receive significant numbers of new residents. By the fourth year, Billings can expect to have received 1854 new households, while Laurel will have grown by 425. These numbers represent percentage increases in population of 3 percent and 9 percent for Billings and Laurel, respectively. The fact that these increases are relatively modest and occur over a 3-year period supports the conclusion that no major deficiencies in services will result.

4. The only in-migration required would perhaps be for engineers and supervisory personnel. In all instances where no in-migration is projected, allowances must be made for the possibility that technical and managerial personnel may be required to relocate by the site.

5. Although it appears low relative to counties in Illinois and West Virginia, the density figure for Yellowstone County is the highest in the state. The figures for population per square mile in Custer and Rosebud counties are 3 and 1, respectively.

6. The multiplier is 1.6, which means that total employment will increase by 1.6 ° 355, or that the increase in secondary employment will be .6 ° 355 = 210.

7. These projections are presented in Stenehjem, E. J., and J. E. Metzer, *A Framework for Projecting Employment and Population Changes Accompanying Energy Development, Phase I*, Draft ANL Report, August 1976. The projections were made using 1970 census figures as a base and adding unemployed and underemployed. In addition, the labor force participation rates were increased to the national percentages in all counties and the increased work force added to the local labor pool. Finally, annual projections of additions to and subtractions from the population between the ages of 16 and 65 have been made. The number of males and females reaching this threshold was multiplied by the labor force participation rate for that cohort and added to the available work force, while those exiting at 65 were subtracted.

Part 3
Reclamation Research in Arid Regions

7

Potentials and Predictions Concerning Reclamation of Semiarid Mined Lands

Richard L. Hodder

Montana Agricultural Experiment Station
Montana State University
Bozeman, Montana

Reclamation or rehabilitation of surface-mined lands is the process of returning disturbed lands to topography, productivity, and use capability that conform to a predesignated purpose. Reclamation must be compatible with or supplementary to use and aesthetics of surrounding land and must minimize environmental degradation. Reclamation of specific mine sites must become a legitimate part of broad land-use planning.

The potential for rehabilitation of any surface-mined area is critically site-specific. The rehabilitation of a mine site depends on: existing ecological and physical conditions, reclamation purpose, mining techniques and equipment used, and skill applied in the technological effort. Two mines, side by side, using two different surface mining techniques, may create distinctly different reclamation situations.

The multiple and interrelated conditions involved in reclamation highlight the absolute importance of preplanning investigations. Such research should precede other phases of logical plan development. The semiarid Western coal region contains a resource so incomprehensibly vast that only preplanning can allow intelligent designation of mining sites in order best to assure acceptable reclamation. Although good potential for reclamation may be apparent, it is achievement—coupled with permanence—that counts. The surest guarantee of success is effective preplanning.

Preplanning, then, sets the stage for development. Only by preplanning is it possible to eliminate the mistakes of the past. Only by effective preplanning can the creative potential of reclamation, which is made possible by mining, be realized. Spoils must be considered and used as a resource, and they must be reclaimed to a permanently useful and productive condition.

Preplanning investigations include: range and wildlife inventories, soil surveys, land-use and grazing patterns, watershed and groundwater studies, archeological review, and overburden analyses. These subjects are of concern either directly or indirectly for a sustained reclamation effort.

Preplanning makes most problems of extraction and rehabilitation predictable. It also makes possible realistic estimates of economic and physical expenditures involved, savings to be realized, and meaningful extraction scheduling. Thus, essential coordination of the total operation becomes possible. Obviously, preplanning places the extraction industry first in line to receive the benefits of the reclamation investment by facilitating prediction of mining problems, estimation of both reclamation and extraction costs, development of more effective mining techniques and more efficient handling of materials, increased sales, and development of by-product resources.

The immediate obligation of agriculturally oriented reclamation is to produce a stable and productive surface through revegetation. A long-term objective must be to produce or protect an acceptable groundwater system. These two obligations may or may not be interdependent. I foresee that overburden analysis and groundwater investigations will soon dominate premining investigative activities. This is the way it should be, because overburden analysis and spoils segregation, coupled with topographic design, are means of protecting the integrity of the groundwater system, which will be the crux of reclamation in the near future.

Core sample analysis of overburden provides such important information as: quantity ratios of overburden, rates of weathering, chemical composition and location of toxic soil materials, soil texture, percolation rates, water-holding capacity, and influences on seed germination and plant growth. Skillful use of this information may be determinant in decisions involving the evaluation of the capabilities of a multimillion dollar machine to do the total mining job, or the consideration of a new mining approach conducive to appropriate spoils segregation practices. Either instance may provide savings in excess of the total reclamation investment. Thus, the most effective spoils handling and reclamation technique should be determined prior to the final design of an efficient mining plan.

Spoils segregation is a frightening subject to many and admittedly an expensive one; however, it is surely a requirement of the near future—and of the present. Salvaging of topsoil is a familiar accepted example of overburden segregation. Special handling and disposal of excessively salty overburden materials is not an unusual occurrence. Overburden analyses and three-dimensional mapping to indicate location, quantities, and depth of toxic materials will be essential in order to designate stratigraphic layers of particular importance and concern, which may require additional segregation or dilution. Innovative techniques to prevent contamination of groundwater will include surface topographic drainage design to reduce or eliminate excessive percolation. Selective placement of toxic spoils materials well above the potentially saturated zone, yet substantially below the plant root zone, may become a common approach. Impermeable protective clay caps placed immediately above concentrations of toxic spoils to divert percolating waters from contacting toxicants are being evaluated.

Surface water impoundments on mined land may be restricted or eliminated, depending on textural and percolation characteristics of the spoils. Fine textured spoils may exhibit such slow percolation and lateral flow as to be inconsequential, whereas loose, sandy, or rocky spoils might be cause for real concern.

Surface mining disturbances in the semiarid West presently seldom exceed 150 feet vertically. Depths to reliable sources of groundwater for domestic use vary from shallow coal seam aquifers to deep wells of 1,000 feet or more. Shallow aquifers affected directly by mining are important, and their integrity must be protected; for they are a common source of water for livestock and subirrigated farmlands. Deeper aquifers may provide alternate sources of water.

Subirrigated alluvium commonly extends in drainages of variable widths from low elevations in major river valleys to high mountain meadows in the Rocky Mountain system. On the other hand, much alluvium is not subirrigated and constitutes innumerable acres of low terraces in major as well as minor valleys. General restrictions on surface mining of alluvial bottoms at any elevation may be illogical as well as impossible to define and regulate. However, surface mining of actual subirrigated alluvial valley floors presently under cultivation would disrupt some of the most productive lands in the West. New legislation may protect the major subirrigated, cultivated alluvial farmlands of the Western states.

"Restoration" of mined land in a strict sense is impossible, for cliffs, rocks, and other features cannot be re-created. Rehabilitation toward a higher and better land use, when feasible, must be the goal. The goal of reshaping of mined lands to approximate original topography must be interpreted and tempered within practical limits, taking into consideration the need to control massive erosion and violent runoff, and to increase the productive potential of mined lands, which previously may have been so rough, rocky, or impassable that they were classed as perhaps aesthetically beautiful but economically unproductive. Formerly steep, impassable and rough topography can be moderately reshaped with gradients reduced to a degree that will provide physical stability and be conducive to the use of farm machinery. With slope reduction, runoff may be reduced, infiltration increased, and seedbed preparation greatly improved. Slopes of 5:1 ratio or less may be sufficiently level to be farmed. Such slopes will become common in order to create lands more useful than the premining terrain. Early spring pasture, hay, or supplemental grain are good potential crops for such reclaimed slopes if soils and climate permit. Recreation and wildlife uses may also fit into a high-priority category at many mine sites.

Surface manipulation of unstable soil surfaces will become a common practice because of the many advantages of this process. Benefits include: improved erosion control, increased infiltration, conservation of soil moisture, and protection of seedlings from the harsh semiarid Western climate.

Surface manipulation offers other advantages, such as single-pass seedbed preparation with simultaneous seeding, thus presenting attractive savings. The practice is conducive to broadcast seeding techniques, which are more aesthetically acceptable than drill seeding and more effective in protecting the soil surface because of the dispersion of seed. The surface manipulation process may offer the only economical means of assuring some response to revegetation attempts when critical precipitation varies from the norm. New forms and sizes of soil surface configurations are being tested at many locations.

Topsoiling dramatically increases infiltration rates and accelerates growth of stabilizing cover, but the practice is not without inherent hazards. It requires timely application and adequate protection. Also, topsoil is literally a storehouse of seed—both desirable and undesirable. The type of seed contained in quantity within the topsoil is determined largely by the range condition and prior use of the area from which the soil was salvaged. A soil from an area in good to excellent condition might be expected to contain many desirable perennial species, whereas topsoil from a source in fair or poor condition would contain undesirable, aggressive, and perhaps noxious weed seed. Many of the aggressive annual weed species are capable of growing under conditions not suitable for germination of desirable perennial species; thus, the undesirable species may gain complete dominance to the detriment of desired or intended forage cover. Once weedy species gain dominance, eradication techniques are difficult and costly. It is usually cheaper and more effective to start the seeding process over again in proper timing with other precautionary measures.

One effective means of subduing weed seed germination in topsoil is to plant a temporary stabilizing species—usually an annual grain—to shade out the weed seedlings and simultaneously protect the investment in topsoil, fertilizer, and seed. The temporary stabilizer may be seeded with or without the perennial seeding mixture, depending on factors such as soil moisture reserves and season of the year.

Temporary stabilizing species are usually annuals. If the annual is seeded singly to protect topsoil until optimum seeding time, the perennial seed mixture can be drilled directly into the protective cover of stubble. This minimum tillage method may preserve the mulching effect provided by the temporary stabilizer.

Depending on the purpose of the reclamation at a particular mine site, the species of vegetation planted will vary. One thing is certain—there will be increased demand for the seed of indigenous species as well as naturalized and introduced species. Many seeds of native species are obtainable by hand collection only. Native species are, however, now appearing on the commercial market through seed companies that specialize in mechanical harvesting of native seed. Recently, superior selections of manageable native species have become available through commercial plantings. Such selections as Critana thickspike wheatgrass, Revenue slender wheatgrass, Rosana western wheatgrass, Lodorm green needlegrass, and Goshen prairie sandreed grass, to name a few, offer improved chances of early response and rapid vigorous establishment, which other native species or strains cannot provide. Production of native species will increase, with the emphasis definitely on selections that can be easily managed, conveniently harvested with commercial combines, and cleaned and processed with standard equipment. Selections of native forage with improved characteristics may be destined to replace other familiar naturalized species in the near future.

Rough and cloddy seedbeds created by surface manipulation processes may become accepted as the standard in seedbed preparation practice. Broadcast seeding immediately following a roughening treatment generally provides very acceptable results with important economic and other advantages over present standard drilling methods. There are, however, some disadvantages also inherent to broadcast seeding. Some of the advantages are:

- Rocky, or inconsistent, soil textures do not affect the seeding process.
- Broadcast seeding can be adapted to varied weather conditions.
- Seedbed preparation and seeding may be combined.
- Time consumed is much reduced over drill methods.
- Seeding equipment generally is much less expensive.
- Depth of seeding is variable, thus more conducive to multitype species seeding.
- Drill row rill erosion is minimized.
- Drill row competition is reduced by improved dispersion of seed placement.

- Dispersion patterns characteristic of broadcast seedings are aesthetically more acceptable.

As mentioned, the potential of vegetation grown on spoils will not be judged by short-term growth but rather by a reasonably long period of continuous productivity. Continuous forage or crop production on spoils is dependent upon proper long-term management, as is the case on any type of site. Effective management is thus the key to long-term reclamation, and without it, achievement of satisfactory revegetation will be temporary.

It stands to reason, then, that if successful reclamation must involve long-term revegetation and if extended forage or crop production is directly dependent on effective management, the reclamation purpose designated early in the preplanning stages of mining must involve local input. Thus, the reclamation will realistically and permanently fulfill a local economic need. Otherwise, the management on which the revegetation depends for survival will be lax or nonexistent, and the reclamation achievement will be misused and thus temporary. Clearly, reclamation designs and purposes will only be successful over the long term if they are ecologically sound and fulfill a real economic need as well as meeting the obligations and requirements of governing regulations.

8

Plants and Treatment for Revegetation of Disturbed Sites in the Intermountain Area

Stephen B. Monsen and A. Perry Plummer

Range Scientist and Botanist-Biologist
Intermountain Forest and Range Experiment Stations
Provo, Utah and Boise, Idaho

INTRODUCTION

The vegetation on some Western rangelands and watersheds has been seriously altered by grazing, road construction, logging, and wildfires. Recent strip-mining has also disrupted numerous sites and the vegetation on them, which ranges from alpine and subalpine plant communities to arid desert shrublands. Most disturbed sites require some type of revegetation. Consequently, a number of adapted plants are needed to treat the different problem areas.

Disturbed areas are often so altered that few plants are capable of colonizing the bare surfaces. Some grasses are widely adapted and a number of them have been planted on such sites. Grasses alone, however, do not always furnish an acceptable plant cover. The inclusion of adapted herbs with shrubs and trees is frequently required.

Recent studies indicate that a number of trees, shrubs, and broadleaf herbs are useful in revegetating disturbed wildlands (Hodder et al. 1970; Plummer 1970, 1977). Recently, seed production centers and nurseries have been developed to produce planting stock of various selections of trees and shrubs (Monsen 1975). Techniques have also been developed for growing and planting woody species as container stock. Such plantings have been successful in adverse environments (Ferguson and Monsen 1974). Good stands have also resulted from planting most shrubs and trees by direct seeding or as nursery and wilding stock.

PROBLEMS OF REVEGETATING DISTURBED LANDS

The features inherent to each disturbance must be recognized before plantings are started. Most disturbances exhibit a number of unique conditions that differ from those in undisturbed situations. These features dictate the selection and use of planted stock. Major factors to consider in stabilizing disturbed sites are: altered planting surfaces, soil stabilization, sedimentation and toxic pollutants, control of undesirable plants, forage resources, game cover and habitat, and aesthetic values.

Altered Surfaces

Soils and planting surfaces exposed by road construction, mining, and logging operations may be so changed that few plants can establish and persist on them. Within each individual disturbance, the edaphic conditions—soil depth, fertility, and amounts of toxic materials—vary greatly. Topsoiling, grading, adding fertilizer, or making other soil amendments help but do not always compensate for serious disruptive influences. Variations within each disturbance cause different conditions at the site and require an assortment of plants to vegetate the site fully. For instance, widely varying soil conditions were left in Teton River Canyon following the failure of an earth-filled dam on June 5, 1976 (Fig. 8.1).

Usually the substrata exposed along roadcuts and waste dumps are highly variable, with both physical and chemical differences existing within any single site. Clayton and Arnold (1972) identified important differences in the degree of weathering and fracturing of granitic rocks within the Idaho Batholith. Weathered and highly fractured rocks are physically able to support seeded plants soon after the materials are exposed by road construction. In contrast, hard rock materials break down slowly when exposed, and these surfaces are not capable of supporting plants for a number of years after being unearthed.

Toxic soil minerals and waste products are frequently encountered in treating mine spoils. These problems, which often occur in isolated areas, create widely different and difficult planting conditions (Farmer, Richardson, and Brown 1976). Soil crusting and compaction also seriously impair the establishment and growth of seedings on some mine tailings and on heavily worked soils.

Sedimentation and Pollution Control

Stabilization and protection of bared surfaces are of prime importance in the rehabilitation of many disturbances. Highly erosive soils often create immense problems when bared of existing vegetation. Megahan (1974).

Figure 8.1 Differences in Planting Conditions Evident Within the Teton River Canyon Following the Failure of an Earth-Filled Dam (Bureau of Reclamation photo taken June 5, 1976, by Glade Walker)

reported a 220 percent increase in erosion from exposed logging roads in south-central Idaho. The deluge of erosive materials occurs within one or two years after the roads are constructed. Plants must be established on roadsides within a short time after the sites are exposed to prevent extensive erosion and sedimentation.

Mining activity is often a continuous operation, and erosive materials slowly develop new tailing piles. Plantings must be scheduled to prevent erosion since mined sites and dumps are often left exposed for extended periods.

A variety of planting stock is frequently required to stabilize steep slopes and roadways. Plants that supply a full ground cover within one growing season are needed on many slopes. Deep-rooted trees and shrubs are often needed to help prevent massive slumping, particularly in regions that receive high amounts of precipitation. Usually, treatments that control soil erosion also tend to help prevent contamination and pollution problems.

Control of Undesirable Plants

During road construction, weeds often invade exposed sites. Once established, they are difficult to control and can seriously impair the growth of more desirable plants. Also, they are a source of seeds that may root on or spread to adjacent agricultural lands.

Topsoil that is removed and stockpiled during road construction is often invaded by cheatgrass brome (*Bromus tectorum*). This annual grass often germinates in the early spring and matures an abundant seed crop within two months. The seeds are distributed as the topsoil is spread over the disturbed areas. Similar problems may occur as halogeton (*Halogeton glomeratus*), a poisonous annual, gains entry on disturbed soils. Dalmation toadflax (*Linaria dalmatica*), another undesirable perennial, frequently establishes in waste areas. Some noxious perennial weeds such as Russian centaurea (*Centaurea repens*), pepperweed whitetop (*Cardaria draba*), morning glory (*Convolvulus arvensis*), and Canada thistle (*Cirsium arvense*) sometimes gain a foothold on disturbed sites.

Planting short-lived nurse crops can prevent the invasion of weeds. Plantings should, of course, include perennials that can persist and prevent the encroachment of undesirables. When noxious weeds are first observed, immediate control measures are important.

Providing Forage

Disturbances usually disrupt and reduce the herbage produced for game and livestock. Frequently, the volume of forage that is lost is not as important as its quality and the period in which it would be utilized.

Where possible, sites should be planted with a mixture of species that can produce nutritious herbage throughout the year. Disturbed sites can often be seeded to grasses and other herbaceous plants, but the establishment of good shrub components may be more difficult to accomplish (Plummer 1977; Plummer, Christensen, and Monsen 1968). Shrubs and trees are particularly useful as winter forage and cover for grazing animals.

Animals are often kept from important grazing land when disruptive activities occur. The bare disturbed sites are usually slow to recover. During this time, forage may be inadequate to sustain grazing animals. Consequently, areas should be immediately planted with species that can quickly supply the needed herbage. Annuals or short-lived perennials can often furnish a rapidly developing forage cover.

Game Cover

Game animals use plantings that provide physical cover and conceal-ment more often than sites that support only a low stand of herbs. Upright shrubs and trees provide escape routes, bedding areas, or calving sites. Variation in cover is required by big game, small mammals, and birds. These resources should be evaluated before the rehabilitation work is begun.

Aesthetic Improvements

Disturbances often alter and reshape the terrain, leaving sites that are aesthetically incompatible with the surrounding landscape. These disturb-ances cannot always be restored or screened through planting, but most areas can be successfully treated by introducing adapted plants that blend with the landscape. Sites can also be reshaped and graded to alleviate some problems.

AMELIORATION OF WILDLAND DISTURBANCES

Plant Selection Trials

Planting a mixture of species expedites the natural sequence of plant succession or development. Recent studies have identified a number of plants useful for initial treatment of disturbed wildlands. Particular qualities in some plants have been detected and have led to the use of various species not formerly considered. Of interest are plants capable of becoming established on exposed infertile sites. These species can initiate changes in the succession of plants, which ultimately produce a stable vegetative composition.

Grasses and Broadleaf Herbs

Annual and biennial herbs have been planted with short-lived perennials on some wildlands with advantageous results. Annual broadleaf herbs have proved highly important in stabilizing erosion. These plants frequently invade disturbances and can create weed problems if left unattended. Annual mustards (*Sisymbrium* spp), Russian thistle (*Salsola kali tenuifolia*), and prickly lettuce (*Lactuca serriola*) invaded the bare canyon slopes created by the failure of the Teton Dam in eastern Idaho. These annual plants furnished a protective ground cover within a short growing period on highly erodible soils. The protected slopes were then seeded to more persistent perennials.

Plants that provide an initial vegetative cover can be classified as (1) pioneer or (2) nurse crops.

Pioneer species: This group includes plants that are not only adapted to grow on severely disturbed sites but are capable of invading such sites to become established. Usually, these plants have vigorous seedlings, produce an abundance of highly viable seed, or possess other characteristics that promote their rapid development. Plants from this group are commonly recommended for treating harsh sites. Only a few species may be adapted to grow on a particular disturbance but, once established, they may persist for many years.

Plants that naturally encroach on disturbed sites can be used as guides in selecting species that will adapt to adverse conditions. A survey of natural disturbances reveals many species that occur in the initial stages of plant community development. Plants that initially appear on old abandoned dredge mines in south-central Idaho are well adapted to rocky, infertile soils. Western yarrow (*Achillea millefolium lanulosa*), low goldenrod (*Solidago multiradiata*), Louisiana sagebrush (*Artemisia ludoviciana ludoviciana*), and Pacific aster (*Aster chilensis adscendens*) have quickly invaded these mine spoils. They have been used in planting similar disturbances. All were readily propagated, and they are now being used in treating harsh conditions. Some native and introduced plants for pioneering situations are listed in Table 8.1. We will briefly describe two of the best ones, western yarrow and mountain rye (*Secale montanum*).

Two related subspecies of western yarrow are found in the West. The subspecies *lanulosa* is restricted to lowlands and is more robust and aggressive than the high elevation subspecies, *alpicola*. As expected, the lowland collections are best suited for arid conditions. Western yarrow produces an abundance of viable seed, and the rapid growth rate of seedlings contributes to its aggressive spreading capabilities. It is moderately palatable and is adapted to a wide variety of sites. Western yarrow persists and increases under harsh conditions; plantings in southern

Table 8.1   Species Recommended for Initial Cover Plantings, with Areas of Adaptation

Common Name	Scientific Name	Short Grass Prairie	Mtn. Brush	Juniper-Pinyon	Big Sage	Northern Desert Shrub	Southern Desert Shrub	Salt Desert Shrub
GRASSES								
Bluegrass, bulbous	Poa bulbosa							
Brome, smooth (southern)	Bromus inermis							
Dropseed, sand	Sporobolus cryptandrus							
Fescue, hardsheep	Festuca ovina duriscula							
Orchardgrass	Dactylis glomerata							
Ricegrass, Indian	Oryzopsis hymenoides							
Rye, mountain	Secale montanum							
Squirreltail, bottlebrush	Sitanion hystrix							
Wheatgrass, crested (Fairway)	Agronyron cristatum							
Wheatgrass, crested (Standard)	A. desertorum							
Wheatgrass, intermediate	A. intermedium							
Wheatgrass, pubescent	A. trichophorum							
Wheatgrass, streambank	A. riparium							
Wheatgrass, tall	A. elongatum							
Wildrye, basin	Elymus cinereus							
FORBS								
Alfalfa, range types	Medicago sativa							
Aster, Pacific	Aster chilensis adscendens							
Burnett, small	Sanguisorba minor							
Cinquefoil, gland	Potentilla glandulosa glandulosa							
Fireweed	Epilobium angustifolium							
Flax, Lewis	Linum lewisii							
Globemallow, gooseberryleaf	Sphaeralcea grossulariaefolia							
Goldenrod, low	Solidago multiradiata							
Goldenweed	Haplopappus spp.							
Lomatium, nineleaf	Lomatium triternatum							
Penstemon, Palmer	Penstemon palmeri							
Sagebrush, Louisiana	Artemisia ludoviciana ludoviciana							
Sagebrush, terragon	A. dracunculus							
Salsify, vegetable-oyster	Tragopogon porrifolius							
Sunflower, common	Helianthus annuus							
Sweetclover, white	Melilotus alba							
Yarrow, western	Achillea millefolium lanulosa							

Idaho have survived for over 45 years and have spread even in unusually dry years. The plants are not susceptible to damaging insect attacks and survive when planted alone on bare clearings. Small transplanted sprigs or root segments of western yarrow have been highly successful in providing a rapidly developing cover on unstable soils (Fig. 8.2). Planting about one pound of seed that has 50 percent purity per acre (0.18 kilogram/hectare [kg/ha]) with other herbs and grasses normally produces a satisfactory component in a stand.

Mountain rye, a short-lived exotic perennial, is adapted to a wide variety of sites. Plantings have been particularly successful on disturbed strip mines in Montana and Wyoming and on arid roadsides and rangelands in Utah, Idaho, and Nevada. Seedlings develop very rapidly and produce an effective ground cover during the year of planting (Fig. 8.3). An abundance of highly palatable herbage is produced, yet the plant is not a serious competitor of slower developing species. Various collections are now under study; growth form, rate of growth, and persistence differ considerably. This grass grows quickly and covers infertile soils. It spreads well naturally and remains as an important element of the vegetative composition. Plantings of mountain rye established over 45 years ago on dry slopes and canyon roadsides in the sagebrush and mountain brush zones of southern Idaho and Utah have persisted and have spread onto a variety of sites.

Other palatable introduced grasses are particularly well suited for planting disturbed sites. The adaptive qualities of Fairway wheatgrass and desert wheatgrass (Agropyron cristatum and A. desertorum) are well known. Desert wheatgrass, recently introduced from the Agricultural Research Service (P.I. 109012), has excellent adaptation to harsh sites. These grasses provide important cover on mine spoils and roadsides that receive 8 to 12 inches of annual precipitation. Intermediate wheatgrass (A. intermedium), pubescent wheatgrass (A. trichophorum), and smooth brome (Bromus inermis) are important sod-formers. They provide excellent cover and forage to disturbed areas where the average annual precipitation is between 12 and 20 inches.

Recently a few specially adapted grasses have been discovered for planting on disturbed lands. Important among these are intermediate wheatgrass selections from the U.S.S.R. (P.I. 314054, 315067, and 315353). These plants produce stronger, more vigorous seedlings and more herbage the first few years after planting than commercial varieties. An impressive drought-tolerant collection of orchardgrass (Dactylis glomerata) (P.I. 109072), developed for rangelands in Utah and Idaho, shows merit. It is well adapted to extremely harsh sites that receive 10 to 12 inches of moisture annually. This same grass selection also grows well and persists on more mesic sites, where it produces an abundance of forage. The plant also

Figure 8.2   Western Yarrow, Easily Established by Direct Seeding or Transplanting onto Disturbed Sites

Figure 8.3   Mountain Rye, a Perennial Used to Suppress the Invasion of Cheatgrass Brome and Other Weeds on Disturbed Rangelands

possesses good regrowth qualities even following prolonged drought. Game animals use it as forage selectively, particularly in the spring and fall months when its green shoots appear before the new growth of other plants.

Four native bunchgrasses well suited to harsh sites are Basin wildrye (*Elymus cinereus*), mountain brome (*Bromus carinatus*), bottlebrush squirreltail (*Sitanion hystrix*), and Indian ricegrass (*Oryzopsis hymenoides hymenoides*). All are readily established by seeding. They are competitive against invasion by undesirable weeds and grow under adverse conditions. As other less adaptive species weaken and die, these grasses increase and spread by natural seed dispersal. All are relatively site-specific, so seed sources should be collected from areas similar to sites on which they are to be planted.

Because of indurated seed coat, Indian ricegrass seeds require a light to medium scarification with a weak solution of sulfuric acid. This process improves seedling establishment (Plummer and Frischknecht 1952).

Fireweed (*Epilobium angustifolium*), gland cinquefoil (*Potentilla glandulosa glandulosa*), low goldenrod, nineleaf lomatium (*Lomatium triternatum*), and Nuttall lomatium (*L. nuttallii*) are an aggressive group of forbs suited to mountain brush communities and to clearings in forest sites. These species are adapted to a range of site conditions that are encountered along roadsides, burns, and other disturbances. They can be seeded in mixtures with grasses.

Nurse crops: Some species not only provide fairly rapid ground cover and forage, but also serve as nurse crops to other desirable plants. The presence and stabilizing influence of these plants furnish better growing conditions for others. Some sites are so harsh that the establishment of a small number of plants can greatly improve the potential for establishment of others. The establishment of Antelope bitterbrush (*Purshia tridentata*) and sulfur eriogonum (*Eriogonum umbellatum*) on rocky dredge mine sites has improved the surface microenvironment. This change has aided in the establishment of other plants. Years may be required before successional change is complete. We recommend, however, planting a mixture of species that establish and develop quickly but do not suppress other long-lived species. Bottlebrush squirreltail, Basin wildrye, mountain rye, and gooseberryleaf globemallow (*Sphaeralcea grossulariaefolia*) can be used as nurse crops on arid rangelands.

Planting Canada bluegrass (*Poa compressa*) on rocky roadsides and mine tailings in northern Idaho has materially aided in the establishment of other plants. This introduced grass is a useful nurse crop, particularly to seeded shrubs. It developed a satisfactory ground cover in the first few years after planting. It also has the unusual ability to remain vigorous when applied fertilizers are depleted and other more sensitive grasses have declined (Fig. 8.4). Canada bluegrass is not seriously competitive to tree and shrub seedlings. Plantings of bottlebrush squirreltail, Indian ricegrass, and sand dropseed (*Sporobolus cryptandrus*) show similar characteristics on more arid sites. Mountain brome can be used on more mesic sites, to which it is well adapted.

Shrubs and Trees

Several shrubs and trees are suitable for direct seeding with forbs and grasses. Seedlings of most woody species develop at a slower rate than do most grasses and herbs and may be suppressed or eliminated by the more rapidly developing herbaceous cover. Shrub and tree seedlings are also more likely to succumb to drought and harsh climatic conditions.

Some shrubs that can be established with seeded herbs are listed in Table 8.2. Rapid development is especially noticeable in fourwing saltbush (*Atriplex canescens*), rubber rabbitbrush (*Chrysothamnus nauseosus*), big sagebrush (*Artemisia tridentata*), and common winterfat (*Ceratoides lanata*) (Table 8.3). Prostrate kochia (*Kochia prostrata*) also grows rapidly on very basic soils.

Planting of shrubs in alternate rows with grasses and forbs increased the survival and growth rate of the woody species in studies conducted in central Utah (Giunta, Christensen, and Monsen 1975). Seeding shrubs in depressions or on protected sites where snow or moisture accumulated also notably improved their success. Machinery has subsequently been developed that creates depressions and eliminates competition for seeded shrubs. Pits left after chaining juniper and piñon trees are good places for establishing shrubs.

Interplanting fourwing saltbush with Fairway crested wheatgrass has markedly improved the forage produced on lands where cheatgrass brome burns occurred in southern Idaho. The addition of Palmer penstemon (*Penstemon palmeri*) has materially aided in the stabilization of road site disturbances on an important portion of a state highway in southern Idaho. The attractive flowers of this species also added aesthetic qualities.

Broadcast and drill seeding of desert bitterbrush (*Purshia glandulosa*), Apache plume (*Fallugia paradoxa*), fringed sagebrush (*Artemisia frigida*), green ephedra (*Ephedra viridis*), and Martin ceanothus (*Ceanothus martinii*) have shown special promise on arid lands in Utah, Nevada, and Idaho.

Broadcast seeding of antelope bitterbrush, Stansbury cliffrose (*Cowania mexicana stansburiana*), blueberry elder (*Sambucus caerulea*), true mountain mahogany (*Cercocarpus montanus*), and Saskatoon serviceberry (*Amelanchier alnifolia*) produced excellent stands when these plants were seeded with grasses on phosphate mining sites in southeastern Idaho.

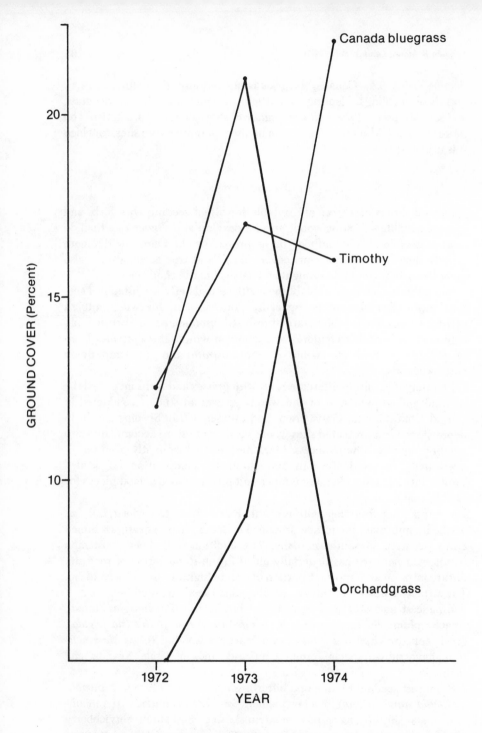

Figure 8.4   Changes in the Percent of Annual Ground Cover of Seeded Grasses as Applied
Fertilizers Are Depleted

TRANSPLANTING TO ESTABLISH WOODY PLANTS

The transplanting of shrubs and trees on disturbed sites has resulted in the successful establishment of several species (Fig. 8.5). Some woody and herbaceous plants that are not easily established by direct seedings can be transplanted to provide an effective cover—e.g., the slowly developing seedlings of Saskatoon serviceberry, squawapple (*Peraphyllum ramosissimum*), and curlleaf mountain mohogany (*Cercocarpus ledifolius*). These shrubs can be planted as container or bare root stock.

Species that germinate erratically produce sporadic stands. If transplants are used, a more reliable cover can be achieved. Nursery stock or rooted cuttings of Woods rose (*Rosa woodsii ultramontana*) and mountain snowberry (*Symphoricarpos oreophilus*) provided useful and adapted material for the raw soils on exposed roadsides in Idaho. These shrubs were easily and quickly established by planting cuttings or nursery stock. Other woody plants, including Scouler willow (*Salix scouleriana*), purple osier willow (*S. purpurea*), skunkbush sumac (*Rhus trilobata trilobata*), oldman wormwood (*Artemisia abrotanum*), and bearberry (*Arctostaphylos uva-ursi*), are easily propagated by stem and root cuttings.

Transplanting nursery or container stock of fast-growing species has been a successful way of stabilizing erodible sites. With this technique, the time required to produce a good ground cover was significantly shortened for snowbrush ceanothus (*Ceanothus velutinus*), Martin ceanothus, and antelope bitterbrush. Shrub transplants have been established in mature stands of grass after small clearings or skelps were made to remove competitive grasses. One-year-old nursery stock can be transplanted onto sites that are newly seeded with grasses and forbs without unusual losses. The establishment of seeded shrubs and trees can be assured if they are seeded a year before grasses are planted.

The planting of container stock rather than bare root plantings of Arizona cypress (*Cypress arizonica*) and mountain mahogany markedly improved the survival of these plants. However, the economics of this technique must be carefully weighed, since bare root plantings are definitely cheaper.

IMPROVING FORAGE RESOURCES WITH SHRUBS AND BROADLEAF HERBS

The addition of shrubs and broadleaf herbs to native grasslands has greatly improved forage resources for livestock and game animals. In one earlier mentioned project, important increases in forage production were achieved through alternate row planting of shrubs and forbs with grasses in

Table 8.2 Attributes of Some Woody Plants for Arid Land Plantings

Common Name	Scientific Name	Establishment by Seed	Seedling Growth Rate	Competitive Ability	Seedling Drought Tolerance
Amorpha, leadplant	Amorpha canescens	4*	5	4	4
Apache plume	Fallugia paradoxa	3	4	4	4
Ash, singleleaf	Fraxinus anomala	3	3	4	4
Bitterbrush, antelope	Purshia tridentata	5	5	4	4
Bitterbrush, desert	P. glandulosa	4	3	4	4
Ceanothus, Martin	Ceanothus martinii	4	4	4	3
Ceanothus, snowbrush	C. velutinus	5	4	5	3
Ceanothus, wedgeleaf	C. cuneatus	3	4	3	3
Cliffrose, Stansbury	Cowania mexicana stansburiana	3	3	3	3
Cypress, Arizona	Cupressus arizonica	2	3	2	2
Elder, blueberry	Sambucus cerulea	3	4	3	3
Ephedra, green	Ephedra viridis	4	4	3	3
Ephedra, Nevada	E. nevadensis	4	3	4	4
Eriogonum, Wyeth	Eriogonum heracleoides	4	4	5	4
Greasewood, black	Sarcobatus vermiculatus vermiculatus	3	3	3	3
Hopsage, spineless	Grayia brandegei	3	2	2	3
Hopsage, spiney	G. spinosa	3	2	2	3
Kochia, prostrate	Kochia prostrata	5	5	4	5
Locust, black	Robinia pseudoacacia	4	4	4	3
Locust, New Mexico	R. neomexicana	3	3	3	4
Peachbrush, desert	Prunus fasciculata	4	4	3	3
Peashrub, Siberian	Caragana arborescens	3	4	3	3
Rabbitbrush, alkali rubber	Chrysothamnus nauseosus consimilis	4	4	5	5
Rabbitbrush, desert	C. stenophyllus	5	3	4	4
Rabbitbrush, low Douglas	C. viscidiflorus viscidiflorus	4	4	4	4
Rabbitbrush, green rubber	C. nauseosus graveolens	4	4	4	4
Russian-olive	Elaeagnus angustifolia	3	3	3	3
Sage, purple	Salvia dorrii carnosa	4	2	2	2

Common Name	Scientific Name	Establish- ment by Seed	Seedling Growth Rate	Compet- itive Ability	Seedling Drought Tolerance
Sagebrush, big	Artemisia tridentata tridentata	4	3	4	4
Sagebrush, bigelow	A. bigelovii	3	3	3	5
Sagebrush, black	A. arbuscula nova	4	3	3	3
Sagebrush, fringe	A. frigida	3	2	3	4
Sagebrush, low	A. arbuscula arbuscula	4	3	3	4
Sagebrush, sand	A. filifolia	4	3	3	4
Sagebrush, spiny	A. spinescens	2	1	2	4
Saltbush, cuneate	Atriplex cuneata	4	3	3	4
Saltbush, fourwing	A. canescens	4	4	4	4
Saltbush, gardner	A. gardneri	3	4	4	4
Saltbush, mat	A. corrugata	1	3	4	4
Serviceberry, Saskatoon	Amelanchier alnifolia	3	2	3	2
Serviceberry, Utah	A. utahensis utahensis	3	2	3	3
Snowberry, longflower	Symphoricarpos longiflorus	3	2	3	3
Snowberry, mountain	S. oreophilus oreophilus	3	2	3	2
Squawapple	Peraphyllum ramosissimum	3	3	3	2
Sumac, Rocky Mtn. smooth	Rhus glabra cismontana	2	3	4	4
Sumac, skunkbush	R. trilobata trilobata	3	3	4	4
Virginsbower, western	Clematis ligusticifolia	4	3	4	4
Winterfat, common	Ceratoides lanta	4	4	4	4

*1=very poor; 2=poor; 3=fair; 4=good; 5=excellent.

Species	Grass cover %	Number of shrubs established (per acre)		
		Maximum	Average	Minimum
Rabbitbrush, rubber	60–70	14,000	7,240	5,100
Rabbitbrush, yellowbrush	40–50	2,400	1,840	700
Sagebrush, big	40–50	1,800	440	160
Saltbush, fourwing	40–50	680	580	270

Table 8.3   Number of Shrubs Established by Direct Seedings with Herbaceous Grasses, Southern Idaho

Figure 8.5   Tree and Shrub Transplants Successfully Established on the Phosphate Mine Disturbances in Southeastern Idaho

northern Utah. Adding fourwing saltbush to grass mixtures markedly increased the yield of herbage for cattle and big game on arid lands in Utah and Idaho. This shrub along with a shorter growing shrub, prostrate kochia, has provided important forage during the fall and winter grazing periods. Only one pound of fourwing saltbush seed per acre (0.18 kg/ha) interplanted with Fairway wheatgrass on a cheatgrass burn in southern Idaho produced about 300 pounds of air-dry forage per acre (55 kg/ha). Between 70 and 80 percent of the annual twig growth has been browsed each year by cattle. However, the plants have persisted and increased in stature over a 10-year period.

Plantings of antelope bitterbrush, Vasey sagebrush (*Artemisia tridentata vaseyana*), and Martin ceanothus are highly productive when seeded on many mined sites. These shrubs are browsed by game animals and furnish useful winter forage and cover.

Inclusion of broadleaf herbs with shrubs and grasses has also improved the forage production and herbage quality of treated sites. Species such as arrowleaf balsamroot (*Balsamorhiza sagittata*) and small burnet (*Sanguisorba minor*) furnish early spring growth that is eagerly sought by both livestock and game. Alfalfa (*Medicago sativa*), Palmer penstemon, and gooseberry globemallow retain considerable green and succulence in the midsummer when other plants are dry. These forbs are particularly adapted to establishment within the big sagebrush types and have proven to be very useful on mine sites.

Russian wildrye (*Elymus junceus*), an introduced grass, supplies midsummer succulence when planted on the salty spoils of mine sites. Although stands are often slow in developing on dry rangelands, this grass establishes fairly quickly on bare mine sites. It is particularly adapted to the infertile substrata that are being exposed by coal mining in many Western states.

Species with Nitrogen-Fixation Capacity

A number of herbs and shrubs possess nitrogen-fixation capabilities, and these can be highly valuable for disturbed lands. The superior ability of some plants to establish on harsh sites is a strong reason for their inclusion in mixtures. Alfalfa, for example, spreads rapidly when seeded on mine spoils and roadsides.

Other nitrogen-fixing herbaceous legumes that are particularly useful on arid sites include Utah sweetvetch (*Hedysarum boreale utahensis*) and various species of lupines (*Lupinus* spp). Plantings of the leguminous shrubs, black locust (*Robinia pseudoacacia*), common bladdersenna (*Colutea arborescens*), and Siberian peashrub (*Caragana arborescens*) have persisted on extremely harsh sites, apparently because of their nitrogen-fixing qualities.

Some nonleguminous species have the capability of producing supplemental amounts of nitrogen through similar microbial associations. Farnsworth and Hammond (1968) report that collections of Louisiana sagebrush have this attribute. This plant has been particularly effective in providing cover on mine sites and roadsides. Various ecotypes of subspecies occur throughout a wide elevational range. To date, the subspecies *ludoviciana* has exhibited the greatest range of adaptation of any collection under test. It has established on acidic soils of the Idaho Batholith and leached mine tailings of the Kennecott Copper Mine near Bingham, Utah. Other forms have been highly successful on alkaline areas.

BREEDING AND SELECTION

Recently, plant breeding programs in Utah have been expanded to produce plants for wildland sites. The plans have taken into account the potential value of a number of native shrubs and forbs. Programs for the selection and breeding of shrubs at the Intermountain Station's Shrub Sciences Laboratory in Provo, Utah, are important for future progress in improving disturbed Western ranges. Particularly promising selections, which have already been mentioned, are rapidly growing shrubs such as fourwing saltbush, big sagebrush, prostrate kochia, and rubber rabbitbrush. When seeded in mixtures, the development of the shrubs is comparable to that of herbaceous species (Plummer 1977).

Recent introductions, supplied by the Agricultural Research Service Laboratory, Logan, Utah, have provided sources of new grasses. Artificial hybridization has resulted in grasses that promise to revegetate disturbed sites in the Snake River Plain of southern Idaho. Hybrids created by crossing *Agropyron libanoticum* with *Elymus canadensis* and *E. multicaulis* with *E. karataviensis* all show particular adaptiveness to the arid conditions. Artificial crosses using bluebunch wheatgrass (*A. spicatum*) and quackgrass (*A. repens*) are providing useful hybrids that are highly productive when grown on dry lowland sites in Idaho. The hybrids, selected for their rooting habits, sustain considerable succulence and green leafy material late into the summer. The retention of green herbage is an effective deterrent to the spread of wildfires.

PLANTING STOCK REQUIRED

Through selection, a number of plants have been identified as being highly successful as range and disturbed land plantings. Yet few of these plants are available in sufficient quantities as planting stock or as seed to be used in large-scale undertakings. Wildland stands do not consistently produce dependable seeds in the quantities necessary to satisfy the needs. When these plants are cultivated, seed production is improved several times in quality and quantity over that of wildland stands. An expanded program for the production of seed and transplant stock is now warranted.

Seed production centers at sites in California, Utah, and Idaho were recently established to produce seed of Lewis flax, Louisiana sagebrush, western yarrow, prostrate kochia, and other plants. Federal, state, and private nurseries have recently been expanded to produce nursery stock and containerized shrubs and trees. Still, more attention must be given to propagating adequate supplies of seeds and transplanting materials for large-scale improvements of disturbed sites. Plants for such improvements have been fairly well identified, but propagation techniques must be developed to supply the urgent demands for planting materials.

REFERENCES

Clayton, James L., and John F. Arnold. 1972. *Practical Grain Size, Fracturing Density, and Weathering Classification of Intrusive Rocks of the Idaho Batholith.* Ogden, Utah: USDA For. Serv. Gen. Tech. Rep. INT–2, Intermt. For. and Range Exp. Stn.

Farmer, E. E., B. Z. Richardson, and R. W. Brown. 1976. *Revegetation of Acid Mining Wastes in Central Idaho.* Ogden, Utah: USDA For. Serv. Res. Pap. INT–178, Intermt. For. and Range Exp. Stn.

Farnsworth, Raymond B., and Max W. Hammond. 1968. "Root Nodules and Isolation of Endophyte on *Artemisia ludoviciana,*" *Utah Acad. Sci., Arts and Letters,* 45(1)182–188.

Ferguson, R. B., and S. B. Monsen. 1974. *Research with Containerized Shrubs and Forbs in Southern Idaho.* N. Am. Containerized For. Tree Seedling Symp., Great Plains Agric. Counc., Publ. 68:349.

Giunta, B. C., D. R. Christensen, and S. B. Monsen. 1975. "Interseeding Shrubs in Cheatgrass with a Browse Seeder-Scalper," *Journal of Range Management,* 28(5):398–402.

Hodder, Richard L., D. E. Ryerson, Ron Mogen, and James Bucholz. 1970. *Coal Mine Spoils Reclamation Research Project.* Prog. Rep. Mont. Agric. Exp. Res. Rep. 8.

Megahan, W. F. 1974. *Deep-Rooted Plants for Erosion Control on Granitic Road Fills in the Idaho Batholith.* Ogden, Utah: USDA For. Serv. Res. Pap. INT–161, Intermt. For. and Range Exp. Stn.

Monsen, Stephen B. 1975. *Selecting Plants to Rehabilitate Disturbed Areas.* Range Manage. Range Symp. Ser. 1, Improved Range Plants.

Plummer, A. Perry. 1970. "Plants for Revegetation of Roadcuts and Other Disturbed Eroded Areas," Ogden, Utah: *Range Improv. Notes,* 15(1):1–8. USDA For. Serv., Intermt. Reg.

Plummer, A. Perry. 1977. "Revegetation of Disturbed Intermountain Area Sites," in J. L. Thames (ed). *Reclamation and Use of Disturbed Land in the Southwest.*

Plummer, A. Perry, and Neil C. Frischknecht. 1952. "Increasing Field Stands of Indian Ricegrass," *Agron. J.,* 44:285–289.

Plummer, A. Perry, D. R. Christensen, and S. B. Monsen. 1968. *Restoring Big Game Range in Utah.* Utah Div. Fish and Game Publ. 68–3.

9

Endomycorrhizae Enhance Shrub Growth and Survival on Mine Spoils[1]

Earl F. Aldon

Principal Hydrologist
Rocky Mountain Forest and Range Experiment Station
U.S. Department of Agriculture
Albuquerque, New Mexico

INTRODUCTION

The term *mycorrhizae* refers to a symbiotic association between plant roots and fungi. There are two distinct kinds, ectocellular and endocellular. This paper will discuss only the endocellular fungi. *Endomycorrhizae,* as they are called, penetrate the walls of the cortical parenchyma and exist in the cell interior. The fungi may penetrate deep into the cortex (Gray 1971). Outside of the cell, the fungal hyphae are thick-walled and irregularly shaped. A loose hyphal net may extend into the soil around the root as far as 1 centimeter (cm) (Gerdemann 1968; Gray 1971).

Some plants derive significant advantages from endomycorrhizal associations. The fungus, for its part, appears to depend on the plant as a carbohydrate source. Mycorrhizae have been shown to be instrumental in increasing phosphate absorption by plants from sources considered to be unavailable (Gray 1971; Gerdemann 1968). Mycorrhizae also appear to be important in protecting plant roots from pathogens. Most of the research on the protective function was done with ectomycorrhizae (Marx 1971; Zak 1964).

Evidence strongly indicates that mycorrhizal fungi produce growth hormones. Horak (1964) showed that mycorrhizae of Norway spruce were able to synthesize indolacetic acid. Davey (1971) reported similar results and indicated that certain species of *Pseudomonas,* harbored by mycorrhizae in the root rhizosphere, also produced indolacetic acid.

174

Endomycorrhizal fungi have not been cultured with much success. A repeatable isolation was conducted by Barret (1961), using a technique in which the VA mycorrhizae were baited with hempseed embryo, but only 2 out of 70 attempts produced pure culture.

Observation of endomycorrhizae is difficult because there is no readily available method for determining their presence other than by staining.

ENDOMYCORRHIZAE ON ARID ZONE SHRUBS

Early Work on Fourwing Saltbush

Survival and growth of fourwing saltbush (*Atriplex canescens* [Pursh] Nutt.) was improved when soil collected from beneath mature plants was added to the potting mix in transplanting experiments. It was hypothesized that the presence of mycorrhizal fungi in this soil was responsible for the increased growth. Greenhouse studies with fourwing saltbush germinated and grown in a soil containing spores of *Glomus mosseae*, a phycomycete reported to form endomycorrhizae on many species of plants, substantiated our hypothesis (Williams, Wollum, and Aldon 1974). Fourwing saltbush grown in soil inoculated with *G. mosseae* were heavier than plants grown in sterilized soil, and were heavily infected with endomycorrhizae.

Field Tests on Mine Spoil

To test these greenhouse findings under field conditions, fourwing saltbush plants were grown from seed in small plant bands by special techniques (Aldon 1975). Soil infested with *G. mosseae* was added to half of the plant bands. Five dewinged fourwing saltbush seeds were planted in each band and covered with additional soil. Seedlings, grown outdoors and irrigated with municipal water, were thinned to one plant per band, overwintered in a lathhouse, and planted the following spring by tested methods on coal mine spoils of the McKinley Coal Mine, 32 kilometers northwest of Gallup, New Mexico. The site is at an elevation of 2100 meters in vegetation of the piñon-juniper-sagebrush type. The spoils material was the result of mining completed four years earlier.

Survival rate was measured in July of the first growing season, and height and diameter of plants and root infection were measured after the second growing season. Controls were found to be nonmycorrhizal, but all treated plants had abundant intracellular and intercellular mycelia.

As Table 9.1 shows, average survival and growth were significantly better on plants grown in soil infested with *G. mosseae*.

	Mycorrhizal	Control
Height (cm)	41.7	27.4
Diameter (cm)	35.8	21.3
Size index (height times diameter)	1493.	584.
Survival (percent)	95.	84.

Table 9.1   Growth and Survival of Mycorrhizal and Nonmycorrhizal Plants

Recent Work with Rubber Rabbitbrush

Seeds of rubber rabbitbrush (*Chrysothamnus nauseosus* [Pall.] Britt.) were placed on the surface of recently graded spoil material in pots and covered with a layer of sterile river sand (Lindsey, Cress, and Aldon 1977). Spoil in half of the pots were inoculated with mycorrhizae obtained from soil collected around rabbitbrush plants growing on undisturbed sites near the spoil banks. Chlamydospores of *Glomus fasciculatus* plus a small number of *G. mosseae* were extracted from 250 cubic centimeters (cc) of soil by the wet sieving and decanting method of Gerdemann and Nicolson (1963). They were used to infest the spoil material in each pot. Plants were grown under greenhouse conditions and watered with tap water.

To evaluate the effect of the mycoflora other than mycorrhizal fungi on plant growth in spoil material, soil washings (in each case 75 milliliters (ml) of the above soil suspension, which passed though a 45 micron (μ) mesh screen) were added to additional pots with spoil material. Rabbitbrush plants (10 plants per treatment) were grown for 132 days in sterilized and nonsterilized mine spoil materials receiving the following treatments: control, soil washings, and mycorrhizal inoculum. Endomycorrhizae were established only in those plants growing in spoil material inoculated with endomycorrhizal fungi. Mycorrhizal plants, growing in both the sterilized and nonsterilized material, had significantly greater dry weights ($> 6x$) and heights ($> 3x$) than those plants without endomycorrhizae. A repeat experiment showed similar results. The mycorrhizal growth response was first observed ten weeks after plants had emerged. In the first and second experiments, approximately 25 percent and 60 percent, respectively, of the nonmycorrhizal plants died after becoming established. In contrast, no mycorrhizal plants died in the first experiment and only one died in the second experiment. *G. fasciculatus* spores were the predominant ones found in the spoil material surrounding the roots of the mycorrhizal plants.

Investigations on Other Plant Species

During the late summer and early fall of 1972, 16 forest and range shrubs from northern New Mexico were examined for endomycorrhizae (Williams and Aldon 1976). Plants selected are important erosion deterrents and are important to grazing and browsing animals. Plants were examined in their natural habitats under a variety of edaphic, climatic, and geographic conditions. Samples of soil and young roots were taken from areas where the population density of the particular plant was high.

Plants with endomycorrhizae were: fourwing saltbush, winterfat (*Eurotia lanata*), mountain serviceberry (*Amelanchier oreophilus*), Utah serviceberry (*Amelanchier utahensis*), true mountain mahogany (*Cercocarpus montanus*), Apache plume (*Fallugia paradoxa*), bush rockspirea (*Holodiscus dumosus*), antelope bitterbrush (*Purshia tridentata*), cliff Fendlerbush (*Fendlera rupicola*), littleleaf mockorange (*Philadelphus microphyllus*), Gambel oak (*Quercus gambelii*), snowberry (*Symphoricarpos* spp), big sagebrush (*Artemisia tridentata*), and skunkbush sumac (*Rhus trilobata*).

In this initial study, no attempt was made to identify the particular fungi with their host plants. In the fall of 1976, three of these original collection sites were revisited and root and soil samples taken near 7 species of the original 14. Collection and staining methods used were similar in both studies. Spores were identified using the Gerdemann and Trappe classification (1974). Vesicles and fungal hyphae were the only structures observed. *Glomus fasciculatus* was found in every collection (Table 9.2). *G. macrocarpus* was found on half of the species sampled, while *G. mosseae* was only found in association with fourwing saltbush.

Staffeldt and Vogt (1975) have found 28 species of Chihuahuan Desert plants to be mycorrhizal. Observations on the occurrence and concentration of endomycorrhizal fungi were made in southern New Mexico. Heavier infections were found on creosotebush (*Larrea tridentata* [DC.] Cov.), honey mesquite (*Prosopis glandulosa* [Torr.] Cockll.), longleaf ephedra (*Ephedra trifurca* [Torr.]), soaptree yucca (*Yucca elata* [Engelm.]), and cactus (*Opuntia* spp). Species of fungi were not identified.

SUMMARY

The importance of vesicular-arbuscular endomycorrhizae on arid-zone plant communities has only recently been extensively investigated.

Shrubs	Number of plants sampled	Endomycorrhizal species found[a]	100-500μ dia roots found mycorrhizal	Spore count/ 100 cc soil[b]
			%	#
Fourwing saltbush	2	1, 2, 3	94	55
Winterfat	2	2, 3	60	44
Mountainmahogany	2	2, 4, 5	53	39
Skunkbush	3	2, 3, 5	70	36
Gambel oak[c]	2	2	86	31
Mockorange	2	2	56	21
Apacheplume	2	2, 3	63	29

[a] 1 = *Glomus mosseae;* 2 = *G. fasciculatus;* 3 = *G. macrocarpus;*
4 = *Sclerocystis spp.;* 5 = *G. microcarpus*
Identifications was based on mature spores only.

[b] Both mature and immature spores were counted.

[c] Some root tips were also found to be ectomycorrhizal.

Table 9.2   Species of Endomycorrhiza Found on Several Shrubs in New Mexico

Fourteen important Southwestern shrub species were found to form endomycorrhizae under field conditions. Fourwing saltbush and rubber rabbitbrush showed significant growth and yield responses when greenhouse plants were inoculated with a known endomycorrhizal fungus. Fourwing saltbush plants inoculated with *Glomus mosseae* were then planted in a field test on coal mine spoils in western New Mexico. After two growing seasons, survival and growth were significantly better on plants grown in soil infested with *G. mosseae*. Examinations at several locations in New Mexico indicated the endomycorrhizal fungi *G. fasciculatus* and *G. macrocarpus* commonly formed associations with seven shrub species.

NOTE

1. The research reported here is a contribution of the SEAM program. SEAM, an acronym for Surface Environment and Mining, is a Forest Service program to research, develop, and apply technology that will help maintain a quality environment and other surface values while helping to meet the nation's mineral requirements.

REFERENCES

Aldon, Earl F. 1975. *Endomycorrhizae Enhance Survival and Growth of Fourwing Saltbush on Coal Mine Spoils.* Fort Collins, Colorado: USDA For. Serv. Res. Note RM–294, Rocky Mtn. For. and Range Exp. Stn.

Barrett, J. T. 1961. "Isolation, Culture, and Host Relation of the Phycomycetoid Vesicular-Arbuscular Mycorrhizal Endophyte *Rhizophagus*," in *Recent advances in botany* (Proceedings of the Ninth International Botanical Congress, Montreal). Toronto, Canada: Univ. of Toronto Press.

Davey, C. B. 1971. "Nonpathogenic Organisms Associated with Mycorrhizae," in: E. Hacskaylo (ed). *Mycorrhizae.* USDA Misc. Publ. 1189.

Gerdemann, J. W. 1968. "Vesicular Arbuscular Mycorrhizae and Plant Growth," *Ann. Review Phytopathology,* 6:397–418.

Gerdemann, J. W. and T. N. Nicolson. 1963. "Spores of Mycorrhizal *Endogone* Species Extracted from Soil by Wet Sieving and Decanting," *Trans. Brit. Mycol. Soc.,* 46:235–244.

Gerdemann, J. W. and J. M. Trappe. 1974. "The Endogonaceae in the Pacific Northwest," *Mycologia Memoir* No. 5. The New York Botanical Garden, Bronx, New York.

Gray, L. E. 1971. "The Physiology of Vesicular Arbuscular Mycorrhize," in E. Hacskaylo (ed). *Mycorrhizae.*

Horak, E. 1964. "Die Bildung Von IES Derivaten Durch Ectotrophe Mykorrhizapilze," *Phytopath. Z.,* 51:491–515.

Lindsey, D. L., W. A. Cress, and Earl F. Aldon. 1977. *The Effects of Endomycorrhizae on Growth of Rabbitbrush, Fourwing Saltbush and Corn in Coal Mine Spoil Material.* Fort Collins, Colorado: USDA For. Serv. Res. Note RM-343, Rocky Mtn. For. and Range Exp. Stn.

Marx, D. H. 1971. "Ectomycorrhizae as Biological Deterrents to Pathogenic Root Infections," in E. Hacskaylo (ed). *Mycorrhizae.*

Staffeldt, E. E. and Vogt, K. B. 1975. *"Mycorrhizae of Desert Plants."* U.S. IBP Desert Biome Research Memorandum 75–37. in *Reports of 1974 Progress,* Vol. 3, "Process Studies Microbiological," Logan: Ecology Center, Utah State University.

Williams, Stephen E., and Earl F. Aldon. 1976. "Endomycorrhizal (Vesicular Arbuscular) Associations of Some Arid Zone Shrubs," *Southwest Natur.,* 20(4):437–444 (10 Jan. 1976).

Williams, S. E., A. G. Wollum II, and Earl F. Aldon. 1974. "Growth of *Atriplex canescens* (Pursh) Nutt. Improved by Formation of Vesicular-Arbuscular Mycorrhizae," *Soil Sci. Soc. Am. Proc.,* 38:962–965.

Zak, B. 1964. "Role of Mycorrhizae in Root Disease," *Ann. Rev. of Phytopathol.,* 2:377–392.

10

An Evaluation of New Mexico Humate Deposits for Restoration of Mine Spoils[1]

James R. Gosz
Larry Barton
Loren D. Potter

Biology Department
University of New Mexico
Albuquerque, New Mexico

INTRODUCTION

New Mexico is in the enviable position of having large deposits of minerals and strip-mineable coal. As a result of current and anticipated mining of these deposits, land restoration and revegetation is becoming a significant concern. Revegetation in the Southwest presents particular problems because of the arid and often alkaline conditions. Treatments designed to increase the moisture-holding capacity and decrease the alkalinity should enhance the success of revegetation.

In many areas of the Southwest there are deposits of humates, some of which are associated with coal beds. Humate contains organic material similar in a number of ways to that found naturally in soil. Because organic matter in soil has important moisture-holding and acidic properties, our project proposed to evaluate humate deposits in terms of their potential in the revegetation of mine spoils. Specifically, we proposed to: (1) evaluate the microbial and mineral nutrient qualities of humates and their effects on mine spoils, and (2) evaluate the effect of humates on plant growth and quality of forage of both native and introduced plants.

PROCEDURES

Chemistry

The most significant indicator of the value of humate is the content of humic matter. The index most commonly used is the amount of sodium hydroxide (NaOH) extractable humic acid. Generally, one to four grams of pulverized (less than 200 mesh) and oven-dried (60–80°C) sample is treated with 50 milliliters (ml) of 0.1 normal NaOH, stirred thoroughly for 1.5–2.0 hours, and centrifuged to separate particulate matter from the solution. The insoluble material may then be treated with sodium hydroxide several times more until it no longer discolors the NaOH solution. Depending upon the amount of humic acid in the sample, four to ten treatments are usually necessary. The soluble humic matter may be further treated with acid and alcohol to separate the humic, fulvic, and ulmic acids. Our value of humic acid is the total amount of NaOH soluble material.

Chemical analyses of humate as well as sandstones and claystone material were made after digestion with equal volumes of nitric and hydrochloric acid (Anderson 1974). This procedure was used because the metals of interest for this work are not bound into the internal structure of sediment particles but are the metals adsorbed to the surface of particles or are precipitated metals from the sample. One gram of oven-dry material was digested by this procedure and brought up to 100 ml volume. This solution was analyzed by atomic absorption spectrophotometry (Perkin Elmer 306) with either the flame (acetylene-air) or heated graphite furnace (HGA–2100) and deterium background correction on appropriate metals. Arsenic and selenium were analyzed using the arsenic-selenium hydride generator, and organic nitrogen was analyzed by the semimicro Kjeldahl procedure (McKenzie and Wallace 1954). Plant tissue was analyzed following grinding in a Wiley mill to pass a 20 mesh screen and digesting with the nitric-hydrochloric acid procedure.

Revegetation

Studies of the use of humate on revegetation were performed at the Jackpile Uranium Mine near Laguna, New Mexico. While the uranium ore comes from the Jackpile sandstone formation, much of the overburden is Tres Hermanos sandstone (Dakota formation). Observations made at the mine site indicated that, while Jackpile sandstone material was generally devoid of vegetation at places where it formed the surface of the dump, the Tres Hermanos sandstone material supported numerous species (Reynolds, Cwik, and Kelly 1976). For this reason, the Tres Hermanos sandstone material was used in the revegetation experiment.

Boulders of Tres Hermanos sandstone with some shale were crushed and leveled in place by driving bulldozers over the surface of an overburden of a disposal dump. An impactor was then driven over the surface to produce a soil-like texture for the seedbed approximately 6 inches (15 centimeters [cm]) deep. Plots were set up on this material in a randomized complete block design with three replicates of 4 humate application rates (0, 800, 4000, 8000 pounds (lbs)/acre (0, 714, 3569, 7138 kilograms/hectare [kg/ha]). The humic acid content of the humate applied was 3 percent of dry weight. Each replicate plot was 25 feet (ft) by 25 ft (58 meters [m]²) separated by 5-feet-wide buffer strips. A second identical randomized block design was set up for the application of barley-hay mulch. These plots were set out and humate levels applied on May 27, 1976. The impactor was driven over the surface again to mix the humate into the soil surface. A seed mixture of nine native grasses, clover, and fourwing saltbush was drilled into the seedbed during the first week of July at a rate of approximately 12 lbs/acre (10.7 kg/ha). On the 15th of August a barley mulch was spread—1900 lbs/acre (1695 kg/ha)—and crimped into this planted surface on one of the experimental randomized blocks.

In late October of 1976, four one-square-meter subplots were clipped in each plot of the randomized complete block design. The plant material was separated into two groups (barley and an aggregate of planted species), brought back to the laboratory, oven-dried (80°C), weighed, and analyzed for chemical content.

Microbiology

Microorganisms were enumerated using standard dilution and cultivation methods with the number of bacteria determined by growth on Plate Count Agar (Difco) and fungi estimated by colony formation on Sabouraud's Dextrose Agar (Difco). All cultures were incubated aerobically at 25°C. Field samples for microbial analysis were taken from the upper 7 cm of soil from at least two areas of each revegetation plot. Root-associated microbes were estimated by mixing the roots of barley in 50 ml of water for one minute using a Waring blender. Microbial content was expressed as the number of organisms present per gram of dry weight of sample.

TERMINOLOGY AND GENERAL PROPERTIES OF HUMATE

The following is offered in an attempt to clarify the confused terminology relating to humate. In the present sense, humate is the salt of humic acid. Humic acid is the organic material that is soluble in NaOH and then can be precipitated in a concentrated acid solution (pH = 2), thereafter remaining insoluble in alcohol. Ulmic and fulvic acids are by-products of such a procedure. Because the term "humic acid" may also refer collectively to the humic, ulmic, and fulvic acids, "humate" may also refer collectively to the salts of those acids (i.e., humate, ulmate, and fulvate).

Actually, humic acids are organic colloids and behave somewhat like clay minerals even though the nomenclature suggests that they are acids and form true salts. When the cation exchange sites on the humic molecule are filled predominantly with hydrogen ions, the material is considered to be an acid and is named accordingly. When the predominant cations on the exchange sites are other than hydrogen—e.g., calcium (Ca), sodium (Na), aluminum (Al), iron (Fe)—the material becomes a salt and is called a humate (Senn and Kingman 1973).

The term humate (or humus) also is used as a lithologic term in reference to carbonaceous mudstone, claystone, and shale and oxidized low-grade coal, which are rich in humic matter. In this paper, the term humate is used almost exclusively as a lithologic term in reference to brownish, carbonaceous mudstones.

On the basis of chemical analyses, the greatest difference between humate and Tres Hermanos sandstone material is in the concentration of four elements. Calcium is higher in the sandstone material, while nitrogen (N), zinc (Zn), and phosphorus (P) are higher in humate. Greater levels in percentage of ash and cation exchange capacity in humates would be attributed to the organic matter in humate.

The chemical content of humate and the humate application rates allow a calculation of the elements added at these rates (Table 10.1). These quantities represent potentially available nutrients for plant uptake and are in the range of fertilizer application rates.

INFLUENCE OF HUMATE ON SOIL MICROBES

Both mulch and humate application caused measurable differences in soil microbial populations (Table 10.2). In the unmulched area, the bacterial number in plots with humate was one-fourth to one-half that found in the nonhumate controls. The abundance of fungi, however, was greatest when 800 lbs/acre (714 kg/ha) of humate was added; it decreased with greater additions of humate. Inoculation of the soil with organisms in the humate was not considered to be an important factor since the humate contained few organisms (i.e., 1.1×10^5 bacteria and 7.4×10^3 fungi per gram dry weight of humate).

In the mulched area, the density of bacteria remained nearly constant

Element	Humate Application Rate		
	714 kg/ha	3659 kg/ha	7138 kg/ha
Ca	34.1	174.7	340.8
Mg	13.0	66.6	129.9
Na	2.6	13.4	26.1
K	1.6	8.0	15.7
N	16.0	81.8	159.5
P	0.9	4.5	8.9
Zn	0.07	0.34	0.67
Cu	0.004	0.020	0.040
Mn	0.16	0.84	1.64
Fe	0.50	2.56	5.00
As	0.001	0.004	0.007
Se	0.0001	0.0002	0.0003
Cation exchange capacity (meq/m^2)	16.1	82.7	161.3

Table 10.1 Quantities of Elements Added at Different Humate Application Rates (data expressed in g/m^2)

with varying humate application, but it was about one-half of the density of the control. Fungal concentrations were considerably reduced with humate applications over that of the control. Perhaps the most significant feature of the microbial study was the association of microbes with plant roots (Table 10.2). The addition of humate at 800 lbs/acre resulted in the greatest number of both bacteria and fungi associated with the roots. Higher levels of humate application appear to inhibit the proliferation of bacteria and fungi that are root-associated. The inhibitory effect of high concentrations of humate was also observed in laboratory studies; however, when humate was added to nutrient media fungal growth was stimulated. Humate appears to be transported or localized on the surface of fungi, and the effect that this activity would have on the plants or microflora of the root zone is unclear at this time.

Humate Application		Mulched				Unmulched	
		Soil		Associated with barley roots		Soil	
kg/ha	lbs/acre	bacteria ($\times 10^7$)	fungi ($\times 10^4$)	bacteria ($\times 10^7$)	fungi ($\times 10^4$)	bacteria ($\times 10^7$)	fungi ($\times 10^4$)
0	0	2.16	13.00	19	160	4.56	2.00
714	800	1.10	1.03	66	490	1.13	4.15
3569	4000	1.05	5.05	16	130	2.29	3.03
7138	8000	1.00	1.23	8	370	2.20	1.03

Table 10.2 Effect of Humate and Mulch Application on the Abundance of Soil Microbes

INFLUENCE ON HUMATE ON PLANT
GROWTH AND NUTRIENT CONTENT

The dearth of precipitation during July and the first two weeks of August 1976 precluded the germination and growth of any seedlings. During the last two weeks of August, precipitation was sufficient to allow germination and growth of planted species. Barley, from seed in the mulch, germinated and became established about five days prior to germination of the planted native species (E. Kelley, personal communication). Precipitation during September was sufficient to maintain the growth of both barley and planted species until the plots were clipped in October.

The data for above-ground growth of the experimental plots, as well as an analysis of variance, showed that the humate and mulching treatments did not significantly affect plant growth. The growth of barley was significantly greater than that of the planted species, but the application of humate was not responsible for any of the difference. The lack of growth response is difficult to interpret at this time since the added nutrients, organic matter, and cation exchange capacity of humate would be expected to cause some effect. In contrast to the microbial studies, there was no suggestion of an inhibitory effect of humate on plant growth.

Although humate and mulching treatments did not measurably affect plant productivity, it is of value to look at the chemical composition of the plant tissue to ascertain the effect of the treatments on forage quality.

Of the elements analyzed to date, only potassium (K) concentrations in barley tissue were affected by humate application. There was a significant ($P < .05$) decrease of K concentrations in barley in all of the plots with any humate applied, but different quantities of humate applied did not cause significant differences in K concentrations. For the planted species, humate application levels did not influence K concentrations in either the mulched or unmulched plots. We cannot explain the decline of K concentrations in barley resulting from humate application. The relatively high concentrations we found in barley are natural for this species (Altman and Dittmer 1972). The planted species had significantly lower K concentrations than barley, and there were significantly higher K concentrations in planted species on the mulched plots, suggesting that there was more available K on those plots caused by the K content of the mulch. Potassium is easily leached from plant tissue (Tukey 1970; Gosz, Likens, and Bormann 1973), and the summer rains could have made the K in the mulch available for plant uptake.

Of the macronutrients, the largest plant accumulation factor (plant/soil) was found for K—no doubt a result of the high requirement for this element and the added levels in the humate and barley supplements. For micronutrients, Zn and selenium (Se) had the highest values. Although the

Se concentration in the soil (.05 parts per million [ppm]) is lower than is typical for sedimentary material (Anderson et al. 1961), the concentrating activity of plants produced Se levels in plant tissue as high as 3.5 ppm. The ability of plants to extract Se is enhanced by aridity, alkalinity, and calcium carbonate deposits, conditions typical of much of the Southwest. Adding various levels of humate had no apparent effect on this phenomenon. Anderson et al. (1961) also report that levels of 5 ppm or more Se in vegetation should be considered as dangerous when ingested by any animal species over a period of several weeks. At this time the vegetation is not considered dangerous. Our future studies will continue to evaluate the effect of humate on plant growth with particular emphasis on the bioaccumulation of potentially dangerous metals.

SUMMARY

Because of the arid and alkaline nature of soils in the Southwest, studies were made of the influence of humate (a salt of humic acid) and mulch application on revegetation success. The results reported are for the first several months of plant growth during the late summer and fall of 1976. Because of this short time interval, all of the effects of humate additions may not have been measurable.

Chemical analyses of the soil and humate material showed humate to differ primarily in content of organic matter and elements normally high in organic matter. The influence of humate on soil microbe counts has been both stimulatory and inhibitory. Of microbes in the soil, only fungi showed some stimulation at intermediate humate applications (800–4000 lbs/acre). Of the microbes associated with plant roots, both bacteria and fungi showed the highest counts at the 800 lbs/acre humate application. Higher application levels seemed to reduce counts.

Neither mulch nor humate application caused a statistically significant effect on plant growth. The only element in plant tissue that appeared affected by humate application was K in barley tissue, which was significantly decreased by all levels of humate. Although humate application has not influenced the heavy metal contents of plants at this point in the study, the natural ability of plants to accumulate potentially toxic elements is cause for additional study.

NOTE

1. Acknowledgments: Financial support was provided by New Mexico Energy Research and Development. We gratefully acknowledge the aid of personnel of the Anaconda Jackpile Uranium Mine for providing the research site and performing the revegetation and mulching operations. Humate was provided by Farm Guard Products, Albuquerque, New Mexico.

REFERENCES

Anderson, J. 1974. "A Study of the Digestion of Sediment by the HNO_3-H_2SO_4 and the HNO_3-HCl Procedures," *Atomic Absorption Newsletter*, 13:31–32.

Anderson, M. S., H. W. Lakin, K. C. Beeson, F. F. Smith, and E. Thacker. 1961. "Selenium in Agriculture," *Agriculture Handbook* No. 200. Agric. Res. Service, U.S. Dept. of Agric.

Altman, P. L., and D. S. Dittmer (eds). 1972. *Biology Data Book*. Vol. I, 2nd ed. Bethesda, Maryland: Federation of American Societies for Experimental Biology.

Gosz, J. R., G. E. Likens, and F. H. Bormann. 1973. "Nutrient Release from Decomposing Leaf and Branch Litter in the Hubbard Brook Forest, New Hampshire," *Ecol. Monogr.*, 43:173–191.

McKenzie, H. A., and H. S. Wallace. 1954. "The Kjeldahl Determination of Nitrogen: A Critical Study of Digestion Conditions—Temperature, Catalyst and Oxidizing Agent," *Aust. J. Chem.*, 7:55–70.

Reynolds, J. F., M. J. Cwik, and N. E. Kelly. 1976. *Reclamation at Anaconda's Jackpile Uranium Mine*. First Annual General Meeting of the Canadian Land Reclamation Association, University of Guelph, Ontario, Canada. Nov. 23, 1976.

Senn, T. L., and A. R. Kingman. 1973. *A Review of Humus and Humic Acids*. Horticulture Dept., South Carolina Ag. Expt. Sta., Clemson, S.C., Res. Ser. No. 145.

Tukey, H. B., Jr. 1970. "The Leaching of Substances from Plants," *Annu. Rev. Plant Phys.*, 21:305–324.

11

Some Applications of Hydrologic Simulation Models for Design of Surface Mine Topography[1]

Roger E. Smith and David A. Woolhiser

Research Hydraulic Engineers
Agricultural Research Service, U.S. Department of Agriculture
Fort Collins, Colorado

INTRODUCTION

Hydrology is a science developed from the empirical description of water movement in nature. Only recently have advances been made in understanding the various physical processes involved. Now the problem of predicting the hydrologic regime of a watershed disturbed by surface mining provides a unique challenge to hydrologic science.

The emphasis in reclamation of disturbed areas has been on establishing a stable, useful vegetal cover. This emphasis implies that the appropriate vegetal cover will provide an optimum balance between site productivity and the production of undesirable residuals, including sediments, inorganic salts, and trace metals. Hydrologic methods can be used to estimate downstream water quality and quantity changes, and to design the watershed to best use available soil materials and water to produce a vegetal cover and to provide long-term stability of the area. Spoil replacement methods may significantly affect subsurface hydrologic conditions; surface macro- and microtopography and topsoil characteristics may affect surface water runoff and plant-useable water (Howard and Wright 1974). (Soil applied to spoil materials often is a mixture of topsoil and subsoil, but here it will be referred to as "topsoil.")

In the limited space available, we will concentrate on the surface-water hydrology of reclaimed areas. The water quality problem involved is largely particulate pollution—erosion of the bare topsoil as vegetation is

189

being established. Climatic factors (the amount and time distribution of precipitation and the evaporative potential), as well as surface topography, topsoil or spoil erosiveness, and cost of various surface treatments must all be considered in the hydrology of reclaimed land.

SOME SIMULATION ABILITIES AND DATA REQUIREMENTS

Simulation models that are developed from the hydraulics of infiltration, runoff, and erosion processes rather than from empirical descriptions allow us to compare in advance several potential topographies of reclaimed surfaces. If we have models that reflect processes and interactions in nature, we can rank alternate reclamation treatments without the precise parameter values needed to predict the quantitative result of any particular treatment.

Comparison or ranking first requires a useful description of the local area's precipitation pattern. A stochastic model of precipitation time series, based on local records or deduced from nearby records, provides data on occurrence, depth, and intensity of precipitation. No simulation will be more accurate than these input data. Even without good local records, however, knowing the structure of the stochastic model will allow us to reasonably estimate its parameters by comparing it with recorded trends of nearby areas (Woolhiser, Rovey, and Todorovic 1973).

From the precipitation model will come either a design storm, a set of design storms, or a representative precipitation period to simulate hydrologic response, depending on the design strategy employed. Next, runoff from each event must be determined by a physically realistic model of infiltration that is sensitive to soil type, vegetal cover, initial soil conditions, and precipitation rate (Smith and Chery 1973). Realistic representation of porous media flow must also be maintained, if snowmelt from the watershed is a major contributor to runoff, as it is in many mine areas in the Western United States.

Overland water flow is then simulated by a surface hydraulic model, in which surface roughness and rill sizes and density are important. Kinematic wave models (Woolhiser 1975) have been shown to be good models for this process.

Surface erosion is highly dependent on raindrop impact and surface-water flow. Therefore, simulation accuracy depends on the accuracy of the models for precipitation, infiltration, and surface-water flow. The work of Meyer and Wischmeier (1969) and Foster and Meyer (1972) are only two examples of the large amount of research on the mechanics of soil erosion. Recently, Smith (1977) has obtained encouraging results in simulating the combined water and sediment yield of individual events on small watersheds.

Models available for simulation in reclamation planning have been discussed briefly. In addition, several models of varying complexity for the evaporative loss of water provide a means to link a time series of events together and estimate initial watershed conditions. The numerical power of the computer allows us to deal with reasonably complicated topographic geometry in surface-mined watersheds.

In estimating the effects of reclamation of a disturbed watershed, the time variation of the parameters in the above models must be predicted to simulate the changes involved in the (presumed) development of a stable cover of vegetation. Figure 11.1 schematically represents the dynamic erosion conditions during revegetation of a hypothetical watershed. The principal contribution of present simulation capabilities is in comparing alternative surface treatments under cost constraints to determine an expected starting erosion rate ($E[Q_s; t = 0]$). The total amount of topsoil lost during the transient period is also important, however, because it will affect the composition and vigor of the vegetation that will become established. Figure 11.1 shows the expected erosion rate; the actual erosion rate may be very different, depending on the climatic sequence experienced. Possibly, the watershed could have climatic conditions and mild-sloped topography with deep, porous topsoil such that surface treatments would reduce runoff—and therefore erosion losses—to a negligible amount. Usually, however, economical surface treatments must allow for some initial erosion loss. What is an acceptable initial loss is a policy matter beyond hydrology. In many areas with very restricted topsoil depths, more stringent erosion controls may be needed.

Topsoil characteristics, actual climatic patterns, and variations in the response of vegetation to precipitation will determine the length of the establishment period for vegetation. As vegetation density increases during the establishment period, erosion from raindrop impact will be reduced; infiltration will be increased, resulting in more soil water and less runoff; and the surface hydraulic roughness will increase. Clearly, the hydrologist needs to work with plant and soil scientists to estimate the vegetal establishment time, the net expected soil loss, and the soil water balance for plant growth during this crucial period.

AN APPLICATION OF SIMULATION TECHNIQUES

One example will illustrate the potential of simulation in designing topographic treatments for controlling erosion on a watershed, and also will show data needs that constrain present model capabilities. The example taken is from a coal strip-mining area in northwest Colorado with an average watershed slope of 8 percent. We will examine only the effect of terrace spacing in reducing erosion and runoff. A single strip the width

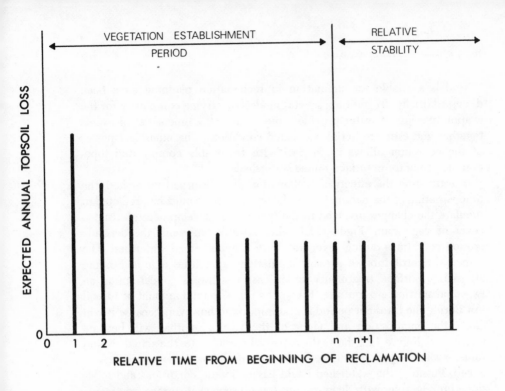

Figure 11.1 Conceptual Model of Expected Changes in Watershed Erosion During Disturbed Land Reclamation

of the watershed is simulated—61 meters (m) (200 feet [ft]) long by 366 m (1200 feet) wide—and two different terrace spacings are compared (Fig. 11.2). The water and sediment discharge hydrographs at point A, resulting from simulating a 2.5 centimeter (cm) (1 inch [in]) storm on this bare soil area for each treatment, are shown in Figure 11.3. This event has approximately a 10-year frequency. The hydraulic response of the two terrace spacings is very similar, but the increased slope length for the 61 m spacing results in almost three times the sediment discharged. In this example, we assumed that for Case I (Figure 11.2) the translation time in the terrace outlet channel from B to A is negligible, with no erosion in the outlet channel. For this example, with terraces of very shallow slope, the actual sediment discharge is a relatively small proportion of the surface erosion, because of deposition in the terraces. Rainfall splash is responsible for considerable amounts of the erosion in this example; and, thus, the sediment is produced early in the runoff period. Such simulations could be used in evaluating terrace spacing, along with information on factors such as (a) the probability of occurrence of this event in the design period, as part of the expected population of runoff events, from which a design input series must be determined; (b) the amount of topsoil that can be lost without detriment to the reclamation objectives; and (c) the construction

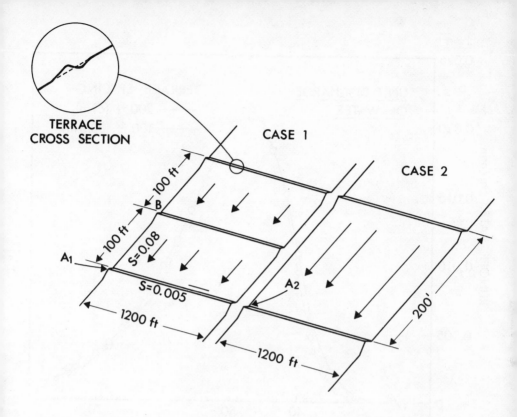

Figure 11.2 A Comparative Terrace Spacing Design Example Problem Employing Hydrologic Erosion Simulation

and maintenance costs for the design terraces. In designing terraces, the overall reclamation strategy and vegetal establishment period must be considered, since it would be deleterious to fill the channels with sediment before a stable vegetal cover is established and, by breaching, promote more serious erosion problems.

Unfortunately, the results illustrated in Figure 11.2 are relative because of present deficiencies in data required for the model. Previous work (Smith 1977) has shown that prediction is good when parameter values are accurately determined, but adequate methodology has not been developed for accurately determining in advance all parameters needed for hydraulics-based infiltration and erosion models.

For relatively uniform-sloped agricultural land with slopes of up to 18 percent and lengths up to 91.5 m (300 ft), an extensive data base is available to aid in predicting mean annual soil loss with the Universal Soil Loss Equation (USLE) of Wischmeier and Smith (1965). Although it was developed for agricultural application, this equation can aid in reclamation design. The USLE is a regression relationship, and because many reclaimed areas have steeper and longer slopes than were included in the

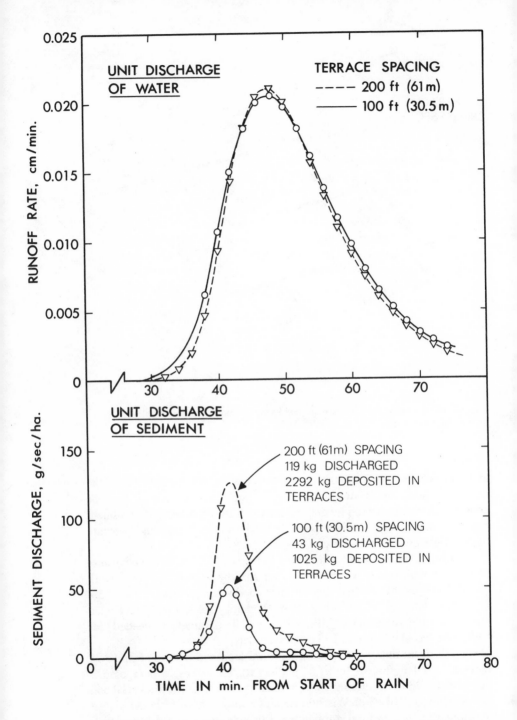

Figure 11.3 Example: Simulated Water and Sediment Discharge Patterns for Two Terrace Spacings on a Uniform Bare Soil Watershed

basic data, it may give misleading results in those cases. The USLE cannot assist in predicting surface runoff.

RESEARCH NEEDS

A more physically based model than the USLE should give better ability to distinguish between the effects of various treatments and topographies, but such a model requires more research on applicable erosion and water carrying capacity relations. Field measurements should be made on the quantitative effects of vegetative cover on reducing raindrop splash erosion, and the effects of vegetal stem density on infiltration rate and hydraulic roughness. A portable field rainfall simulator is being planned to study the time change in parameters at selected sites during the transient stage of vegetative establishment. Well-designed laboratory studies can also contribute basic knowledge to aid in simulation model applications.

For an accurate hydraulic erosion relation, we must develop some method to select reasonable parameters for erosion relations by physical measurement of spoil or topsoil material. In addition, we are not yet able to predict, for a given slope and soil, the expected extent of rill formation. Rills alter the hydraulic surface characteristics drastically and, therefore, affect the erosion and depositional mechanisms.

SUMMARY

With the legal requirements and time constraints on reclamation efforts, we can use hydrologic models in evaluating various reclamation treatments. Empirical data used in the USLE provide guidelines for topographic design, but reclamation problems often will vary considerably from the conditions for which this relation has proved useful. More detailed investigations are needed to establish parameters for hydraulic erosion rates on a range of soils, and physical relations between soil properties and erosion susceptibilities. The hydrologic effects of increasing density of cover must be quantified so that the time required to reach a "stable," or legally reclaimed, watershed may be estimated more easily.

Present capability in simulation is adequate for constructing sequences of statistically representative precipitation for simulation and in comparing the hydraulic response of various possible topographic treatments. For reclaimed areas, the dynamic changes and interrelations of erosion, vegetal establishment, and surface hydraulics offer the greatest challenge to hydrology.

NOTE

1. Acknowledgments: This paper is a contribution from the Agricultural Research Service, USDA, in cooperation with the Colorado State University Experiment Station. This research was supported in part by funds from the U.S. Environmental Protection Agency thru EPA-IAG-D6-E763.

REFERENCES

Foster, G. R., and L. D. Meyer. 1972. "A Closed Form Soil Erosion Equation for Upland Areas," in H. W. Shen (Ed). *Sedimentation,* Fort Collins, Colorado.

Howard, James F., and Richard E. Wright. 1974. "Evaluation Procedure of Critical Factors of Mining Impact on Ground Water Resources," *Water Resources Problems Related to Mining,* Minneapolis: American Water Res. Assoc., Proc. No. 18.

Meyer, L. D., and W. H. Wischmeier. 1969. "Mathematical Simulation of the Process of Soil Erosion by Water," *Trans. ASAE,* 12(6).

Smith, R. E. 1977. "Field Test of a Distributed Watershed Erosion/Sedimentation Model," *Soil Erosion: Prediction and Control,* Soil Conservation Society of America, Special Publication 21, pp. 201–9.

Smith, R. E., and D. L. Chery, Jr. 1973. "Rainfall Excess Model from Soil Water Flow Theory," *Journal of the Hydraulics Division, Proceedings ASCE,* Vol. 99 (HY9).

Wischmeier, W. H., and D. S. Smith. 1965. "Predicting Rainfall-Erosion Losses from Cropland East of the Rocky Mountains," *Agri. Handbook No. 282,* ARS, USDA.

Woolhiser, D. A. 1975. "Simulation of Unsteady Overland Flow," in K. Mahmood and V. Yevjevich (eds). *Unsteady Flow in Open Channels, Vol. II,* Fort Collins, Colorado: Water Resources Publications.

Woolhiser, D. A., E. Rovey, and P. Todorovic. 1973. "Temporal and Spatial Variation of Parameters for the Distribution of N-Day Precipitation," in E. F. Schulz, V. A. Koelzer, K. Mahmood (eds). *Floods and Droughts; Proceedings of the Second International Symposium in Hydrology,* September 11-13, 1972, Fort Collins, Colorado: Water Resources Publications.